T0220114

LONDON MATHEMATICAL SOCIETY STUDENT TEXTS

Managing Editor: Professor D. Benson,
Department of Mathematics, University of Aberdeen, UK

London Mathematical Society Student Texts 81

Number Theory, Fourier Analysis and Geometric Discrepancy

GIANCARLO TRAVAGLINI
Università di Milano-Bicocca

CAMBRIDGE
UNIVERSITY PRESS

University Printing House, Cambridge CB2 8BS, United Kingdom

One Liberty Plaza, 20th Floor, New York, NY 10006, USA

477 Williamstown Road, Port Melbourne, VIC 3207, Australia

314-321, 3rd Floor, Plot 3, Splendor Forum, Jasola District Centre, New Delhi - 110025, India

79 Anson Road, #06-04/06, Singapore 079906

Cambridge University Press is part of the University of Cambridge.

It furthers the University's mission by disseminating knowledge in the pursuit of education, learning and research at the highest international levels of excellence.

www.cambridge.org
Information on this title: www.cambridge.org/9781107619852

© G. Travaglini 2014

First published 2014

A catalogue record for this publication is available from the British Library

Library of Congress Cataloging in Publication data
Travaglini, Giancarlo, author.
Number theory, Fourier analysis and geometric discrepancy / Giancarlo Travaglini.
pages   cm. – (London Mathematical Society student texts)
Includes bibliographical references and index.
ISBN 978-1-107-04403-6 (hardback) – ISBN 978-1-107-61985-2 (paperback)
1. Number theory – Textbooks.   I. Title.
QA241.T68   2014
512.7–dc23
2014004844

ISBN 978-1-107-04403-6 Hardback
ISBN 978-1-107-61985-2 Paperback

To Frances

Contents

Introduction

Through this book we wish to achieve and connect the following three goals:

1) to present some elementary results in number theory;
2) to introduce classical and recent topics on the uniform distribution of infinite sequences and on the discrepancy of finite sequences in several variables;
3) to present a few results in Fourier analysis and use them to prove some of the theorems discussed in the two previous points.

The first part of this book is dedicated to the first goal. The reader will find some topics typically presented in introductory books on number theory: factorization, arithmetic functions and integer points, congruences and cryptography, quadratic reciprocity, and sums of two and four squares. Starting from the first few pages we introduce some simple and captivating findings, such as Chebyshev's theorem and the elementary results for the Gauss circle problem and for the Dirichlet divisor problem, which may lead the reader to a deeper study of number theory, particularly students who are interested in calculus and analysis.

In the second part we start with the uniformly distributed sequences, introduced in 1916 by Weyl and related to the strong law of large numbers and to Kronecker's approximation theorem. Then we introduce the definition of discrepancy, which is the quantitative counterpart of the uniform distribution and has natural applications in the computation of high-dimensional integrals. For the particular case of integer points we use different techniques to prove some classical but not trivial results for the Gauss circle problem and for the Dirichlet divisor problem. Then we introduce the *geometric discrepancy*, also known as *irregularities of distribution* because some of its main results show the existence of unavoidable errors in the approximation of a continuous object by a discrete sampling. This theory has grown over the last 60 years thanks to the contributions by Roth, Schmidt, Beck and other authors, and it is presently a crossroads between number theory, combinatorics, Fourier analysis, algorithms and complexity, probability and numerical analysis [49]. Its current applications range from traditional science and engineering to modern computer science and financial mathematics [43].

A large number of the results in this book are proved through Fourier analytic arguments: pointwise convergence of Fourier series, completeness of the trigonometric system, trigonometric approximation, Poisson summation formula, exponential sums, decay of Fourier transforms and Bessel functions. The result is a short and self-contained course on Fourier analysis, which we present in parallel with the two previous points.

This book is based on a number theory course of 60–70 hours given by the author at the University of Milano-Bicocca for several years. It was a postgraduate course, but many undergraduate students attended it with success. The prerequisites are limited: no prior knowledge of number theory or Fourier analysis is necessary, we assume a bit of algebra and a solid background in calculus, we need the Lebesgue integral, but we use complex analysis only in the last chapter.

The lecture notes of the above course first appeared in the Italian book *Appunti su teoria dei numeri, analisi di Fourier e distribuzione di punti* [173]. We wish to thank the Unione Matematica Italiana who generously released the rights to this revised and expanded English version.

We are very grateful to the students who attended the course. We also wish to thank Anatoly Podkorytov, Giacomo Gigante, Leonardo Colzani, Luca Brandolini and William Chen, who have read the original draft. We thank William Chen also for permission and encouragement to freely use his notes [46] of the number theory courses he taught at Imperial College London. These notes were modified many times from notes used by various colleagues over many years, both at Imperial College London and University College London. It seems likely that Davenport is the original source of these notes.

We are also happy to thank all the people at Cambridge University Press who were involved in publishing this book, in particular Roger Astley, Roisin Munnelly, Samuel Harrison and Joanna Breeze.

Giancarlo Travaglini
March 2014

PART ONE

ELEMENTARY NUMBER THEORY

1

Prelude

Number theory deals with the properties of the positive integers, which were probably the first mathematical objects discovered by human beings. In this chapter we shall initially study the factorization of positive integers into primes, a basic result called the *fundamental theorem of arithmetic*. The possibly exaggerated title 'Prelude' refers to the second section, where we introduce Chebyshev's theorem on the distribution of prime numbers. This result is remarkable and yet rather easy to understand, and it may encourage the reader to approach more advanced topics in number theory.

For the first part of this book we have used various references, including [3, 4, 6, 8, 9, 42, 46, 63, 68, 72, 76, 90, 93, 96, 101, 103, 108, 119, 120, 127, 128, 136, 145, 151, 165].

1.1 Prime numbers and factorization

We shall denote by $\mathbb{N} = \{1, 2, \ldots\}$ the set of natural numbers and by $\mathbb{Z}$ the set of integers. We shall say that $0 \neq b \in \mathbb{Z}$ divides $a \in \mathbb{Z}$ if there exists $c \in \mathbb{Z}$ such that $a = bc$. In this case we shall write $b \mid a$. If b does not divide a we shall write $b \nmid a$.

We know[1] that, given $a \in \mathbb{Z}$ and $b \in \mathbb{N}$, there exist (unique) $q, r \in \mathbb{Z}$ such that $a = bq + r$, with $0 \leq r < b$. We present the following consequence.

Theorem 1.1 *Let $b > 1$ be an integer. Then every $a \in \mathbb{N}$ can be written in one and only one way in base b :*

$$a = c_0 + c_1 b + c_2 b^2 + \ldots + c_n b^n , \tag{1.1}$$

---

[1] The set $\{a - xb : x \in \mathbb{Z}\} \cap (\mathbb{N} \cup \{0\})$ is not empty and let $r = a - qb$ be its minimum. Observe that $(0 \leq)r < b$, otherwise we would have $r = a - qb \geq b$ and $0 \leq a - (q + 1)b < r$, against the minimality of r. Now assume that $a = bq_1 + r_1 = bq_2 + r_2$, then $|r_1 - r_2| = b|q_2 - q_1|$. If $q_1 \neq q_2$ then $|r_1 - r_2| = b|q_2 - q_1| \geq b$, which is impossible since $0 \leq r_1, r_2 < b$.

3

where $n \geq 0$ and $0 \leq c_j < b$ for $j = 0, \ldots, n-1$, while $1 \leq c_n < b$.

Proof Let us prove by induction that if $b^n \leq a < b^{n+1}$, then a can be written in base b. This is true for $n = 0$ and we assume that it is true for every $0 \leq m \leq n-1$. Since $b^n \leq a < b^{n+1}$ we have

$$a = c_n b^n + r \,,$$

with $0 \leq r < b^n$ and $1 \leq c_n < b$. If $r = 0$, then a is written as in (1.1). If $r > 0$, we recall that $r < b^n$ and thus $b^m \leq r < b^{m+1}$ for some $m \leq n-1$. We now use the induction assumption to write

$$r = p_0 + p_1 b + p_2 b^2 + \ldots + p_m b^m$$

with $0 \leq p_j < b$ for $j = 0, \ldots, m$. Then

$$a = p_0 + p_1 b + p_2 b^2 + \ldots + p_m b^m + c_n b^n \,.$$

Finally, we assume that there are two ways to write a in (1.1). By suitably subtracting them we obtain

$$0 = q_0 + q_1 b + q_2 b^2 + \ldots + q_k b^k$$

with $q_k \geq 1$ ($k = 0$ gives a contradiction, so we assume that $k \geq 1$). For every $0 \leq j \leq k-1$ we have $|q_j| \leq b-1$. Then we have

$$q_k b^k = -\left(q_0 + q_1 b + q_2 b^2 + \ldots + q_{k-1} b^{k-1} \right)$$

and

$$b^k \leq q_k b^k \leq |q_0| + |q_1| b + |q_2| b^2 + \ldots + |q_{k-1}| b^{k-1}$$
$$\leq (b-1)\left(1 + b + b^2 + \ldots + b^{k-1} \right) = b^k - 1 \,,$$

which is impossible. $\square$

The following theorem introduces the definition of greatest common divisor.

Theorem 1.2 *Let a, b be two integers not both zero. Then there exists a unique $d \in \mathbb{N}$ such that*

(i) there exist $x, y \in \mathbb{Z}$ satisfying $d = ax + by$,
(ii) $d \mid a$ and $d \mid b$,
(iii) if $k \in \mathbb{N}$ divides a and b, then it also divides d.

Proof Let

$$I := \{au + bv\}_{u,v \in \mathbb{Z}} \,.$$

Then $I \cap \mathbb{N}$ is not empty and let $d = \min(I \cap \mathbb{N})$. Observe that d trivially satisfies (i) and (iii). In order to show that d satisfies (ii), it is enough to prove that d divides every element in I. Indeed, let $au + bv = z \in I$, assume that $q \in \mathbb{Z}$ and $0 \le r < d$ satisfy $z = dq + r$. Then

$$r = z - dq = au + bv - (ax + by)q = a(u - xq) + b(v - yq) \in I .$$

Since $0 \le r < d = \min(I \cap \mathbb{N})$ we deduce that $r = 0$, thus d divides z. The uniqueness follows from (ii) and (iii). □

The number d is called the *greatest common divisor* (gcd) of a and b. We shall write[2]

$$d = (a, b) .$$

When $(a, b) = 1$ we shall say that a and b are coprime. Observe that $(a, b) = 1$ if and only if there exist integers x, y such that $ax + by = 1$.

Theorem 1.3 (Euclid's lemma)   *Let a, b, c satisfy $a \mid bc$ and $(a, b) = 1$. Then $a \mid c$.*

Proof   Let x and y satisfy $ax + by = 1$. Then $c = cax + cby$. Since $a \mid acx$ and $a \mid bcy$, we deduce that $a \mid c$. □

We are going to describe a famous method, called the *Euclidean algorithm*, which gives the gcd of two positive integers. We need a lemma.

Lemma 1.4   *Let $a, b, q, r \in \mathbb{N}$ satisfy $a = qb + r$. Then $(a, b) = (b, r)$.*

Proof   If $t \mid b$ and $t \mid r$, then $t \mid a$. In particular, $(b, r) \mid a$. Since $(b, r) \mid b$ we deduce that $(b, r) \mid (a, b)$. In the same way we see that $(a, b) \mid (b, r)$. □

The Euclidean algorithm uses the previous lemma to compute the gcd of two integers $a \ge b > 0$. Indeed, let us write

$$\begin{aligned}
a &= q_1 b + r_1 && \text{with } 0 < r_1 < b \\
b &= q_2 r_1 + r_2 && \text{with } 0 < r_2 < r_1 \\
r_1 &= q_3 r_2 + r_3 && \text{with } 0 < r_3 < r_2 \\
r_2 &= q_4 r_3 + r_4 && \text{with } 0 < r_4 < r_3
\end{aligned}$$

$$\vdots$$

$$\begin{aligned}
r_{n-3} &= q_{n-1} r_{n-2} + r_{n-1} && \text{with } 0 < r_{n-1} < r_{n-2} \\
r_{n-2} &= q_n r_{n-1} .
\end{aligned}$$

By the previous lemma we have

$$(a, b) = (b, r_1) = (r_1, r_2) = (r_2, r_3) = \ldots = (r_{n-3}, r_{n-2}) = (r_{n-2}, r_{n-1}) = r_{n-1} .$$

---

[2] The symbol (a, b) already denotes the open interval $(a, b) = \{x \in \mathbb{R} : a < x < b\}$ and the pair $(a, b) \in \mathbb{Z}^2$, but there is actually no confusion.

Then the Euclidean algorithm says that $(a, b) = r_{n-1}$.

We apply the Euclidean algorithm to compute $(35777, 4123)$. We have

$$35777 = 4123 \cdot 8 + 2793 \qquad (1.2)$$
$$4123 = 2793 \cdot 1 + 1330$$
$$2793 = 1330 \cdot 2 + 133$$
$$1330 = 133 \cdot 10 .$$

Hence $(35777, 4123) = 133$.

Remark 1.5 If we look at the numbers in the algorithm we can find x, y in Theorem 1.2. Indeed, if we rewrite the lines in (1.2) with $a = 35777$ and $b = 4123$ we have

$$a = 8b + 2793$$
$$b = (a - 8b) \cdot 1 + 1330$$
$$a - 8b = [b - (a - 8b) \cdot 1] \cdot 2 + 133 .$$

Then we have

$$133 = 3a - 26b$$

or

$$(35777, 4123) = 3 \cdot 35777 - 26 \cdot 4123 .$$

We now introduce the prime numbers.

Definition 1.6 An integer $p > 1$ is a *prime number* if it has only two positive divisors (namely 1 and p). We shall write $\mathcal{P}$ for the set of prime numbers. We shall say that an integer $n > 1$ is a *composite number* if $n \notin \mathcal{P}$.

Lemma 1.7 *Let $a, b \in \mathbb{N}$ and let p be a prime. If $p \mid ab$, then $p \mid a$ or $p \mid b$.*

Proof $p \in \mathcal{P}$ and $p \nmid a$ imply $(a, p) = 1$. Then Theorem 1.3 gives $p \mid b$. □

By applying the previous lemma several times we obtain the following result.

Lemma 1.8 *Let $a_1, a_2, \ldots, a_k \in \mathbb{N}$ and let p be a prime. If $p \mid (a_1 a_2 \cdots a_k)$, then $p \mid a_j$ for some $j = 1, \ldots, k$.*

We can now introduce the *fundamental theorem of arithmetic*.

Theorem 1.9 (Fundamental theorem of arithmetic)   *Every integer $n > 1$ can be written in a unique way (up to permutation) as*

$$n = p_1^{m_1} p_2^{m_2} \cdots p_N^{m_N}, \tag{1.3}$$

where the numbers p_j are prime and the numbers m_j are positive integers.

In certain cases it may be useful to write $n = p_1^{m_1} p_2^{m_2} \cdots p_N^{m_N} \cdots$ as an infinite product, where $\{p_j\}_{j=1}^{+\infty} = \mathcal{P}$ and all but a finite number of exponents m_j are zero. (1.3) is called the *canonical decomposition* (or *factorization*) of n.

Proof   We shall use induction to prove that every integer $n \geq 2$ can be written as a product of prime numbers. This is true for 2 and we consider $n > 2$. Assume that every $2 \leq m \leq n - 1$ can be written as a product of prime numbers. If $n \in \mathcal{P}$, we are done. If not, let $n = n_1 n_2$. Then, by the induction assumption, n_1 and n_2 are products of prime numbers. Then the same is true for n. Now we shall prove the uniqueness of the decomposition. Let $\mathcal{A} \subset \mathbb{N}$ be the set of natural numbers with more than one canonical decomposition. Let $M = \min \mathcal{A}$. Then we may write

$$M = p_1 p_2 \cdots p_r = p_1' p_2' \cdots p_s',$$

where $p_1, p_2, \ldots, p_r$ and $p_1', p_2', \ldots, p_s'$ are prime numbers (possibly repeated), and the two products differ for at least one term. By Lemma 1.8 we deduce that $p_1 = p_j'$ for a suitable j. Then

$$\widetilde{M} = p_2 \cdots p_r = p_1' \cdots p_{j-1}' p_{j+1}' \cdots p_s'$$

is smaller than M and admits two different canonical decompositions.   □

See [20, 6.1] for a comment on the above theorem and its history. See also [61, Ch. 8] and [85].

The fundamental theorem of arithmetic allows us to write the gcd as follows. Let $a, b \in \mathbb{N}$. We may write $a = p_1^{\alpha_1} p_2^{\alpha_2} \cdots p_s^{\alpha_s}$ and $b = p_1^{\beta_1} p_2^{\beta_2} \cdots p_s^{\beta_s}$ (with the same prime numbers in the two products) as long as we allow some of the exponents to be zero. Then

$$(a, b) = p_1^{\min(\alpha_1, \beta_1)} p_2^{\min(\alpha_2, \beta_2)} \cdots p_s^{\min(\alpha_s, \beta_s)}.$$

Observe that the Euclidean algorithm does not need the fundamental theorem of arithmetic.

Euclid has proved the infinitude of primes.

Theorem 1.10 (Euclid)   *There are infinitely many prime numbers.*

Proof   Assume that $\mathcal{P} = \{p_1, p_2, \ldots, p_N\}$. Then every other number should be a product of elements in $\mathcal{P}$. But this is impossible for $(p_1 p_2 \cdots p_N) + 1$.   □

Here we can see the original proof.

'Prime numbers are more than any assigned multitude of prime numbers. Let A, B, and C be the assigned prime numbers. I say that there are more prime numbers than A, B, and C. Take the least number DE measured by A, B, and C. Add the unit DF to DE. Then EF is either prime or not. First, let it be prime. Then the prime numbers A, B, C, and EF have been found which are more than A, B, and C. Next, let EF not be prime. Therefore it is measured by some prime number. Let it be measured by the prime number G. I say that G is not the same with any of the numbers A, B, and C. If possible, let it be so. Now A, B, and C measure DE, therefore G also measures DE. But it also measures EF. Therefore G, being a number, measures the remainder, the unit DF, which is absurd. Therefore G is not the same with any one of the numbers A, B, and C. And by hypothesis it is prime. Therefore the prime numbers A, B, C, and G have been found which are more than the assigned multitude of A, B, and C. Therefore, prime numbers are more than any assigned multitude of prime numbers.'

(Euclid, *Elements*, Book IX)

Observe that the above argument does not offer an instrument to find prime numbers. Indeed,

$$(2 \cdot 3 \cdot 5 \cdot 7 \cdot 11 \cdot 13) + 1 = 30031 = 59 \cdot 509$$

is a composite number.

The following result was proved by Euler in 1737.

Theorem 1.11 (Euler)

$$\sum_{p \in \mathcal{P}}^{+\infty} \frac{1}{p} = +\infty . \tag{1.4}$$

First proof   For every real number $x \geq 2$ we write

$$P_x := \prod_{\mathcal{P} \ni p \leq x} \left(1 - \frac{1}{p}\right)^{-1} .$$

Since

$$\log(1 + \varepsilon) = \varepsilon - \frac{1}{2}\varepsilon^2 + \frac{1}{3}\varepsilon^3 - \frac{1}{4}\varepsilon^4 + \ldots$$

for $-1 < \varepsilon < 1$, we have

$$\log P_x = -\sum_{\mathcal{P} \ni p \leq x} \log\left(1 - \frac{1}{p}\right) = \sum_{\mathcal{P} \ni p \leq x} \left(\frac{1}{p} + \frac{1}{2p^2} + \frac{1}{3p^3} + \frac{1}{4p^4} + \ldots\right)$$

$$\leq \sum_{\mathcal{P} \ni p \leq x} \frac{1}{p} \left(1 + \frac{1}{2^2} + \frac{1}{2^3} + \frac{1}{2^4} + \ldots \right) = \frac{3}{2} \sum_{\mathcal{P} \ni p \leq x} \frac{1}{p}.$$

To complete the proof we show that $P_x \to +\infty$ when $x \to +\infty$. Indeed, let $p_1 < p_2 < p_3 < \ldots < p_N$ be the prime numbers $\leq x$. Since

$$\left(1 - \frac{1}{p} \right)^{-1} = \sum_{j=0}^{+\infty} p^{-j}$$

we have

$$P_x = \prod_{\mathcal{P} \ni p \leq x} \left(1 - \frac{1}{p} \right)^{-1} \tag{1.5}$$

$$= \prod_{j=1}^{N} \left(1 - \frac{1}{p_j} \right)^{-1}$$

$$= \left(1 + \frac{1}{p_1} + \frac{1}{p_1^2} + \ldots \right) \left(1 + \frac{1}{p_2} + \frac{1}{p_2^2} + \ldots \right) \cdots \left(1 + \frac{1}{p_N} + \frac{1}{p_N^2} + \ldots \right)$$

$$= 1 + \left(\frac{1}{p_1} + \frac{1}{p_2} + \ldots + \frac{1}{p_N} \right) + \left(\frac{1}{p_1^2} + \frac{1}{p_1 p_2} + \frac{1}{p_2^2} + \frac{1}{p_1 p_3} + \ldots \right)$$

$$+ \left(\frac{1}{p_1^3} + \frac{1}{p_1^2 p_2} + \ldots \right) + \ldots$$

$$= \sum_{\substack{n = p_1^{m_1} p_2^{m_2} \cdots p_N^{m_N}, \\ p_1 \leq x, \, p_2 \leq x, \, \ldots \, p_N \leq x}} \frac{1}{n}$$

$$\geq \sum_{k \leq x} \frac{1}{k} \longrightarrow +\infty$$

as $x \to +\infty$. $\qquad\qquad\qquad\qquad\qquad\qquad\qquad\qquad\qquad\qquad\square$

An argument similar to the one in (1.5) shows that for every real number $s > 1$ we have

$$\sum_{n=1}^{+\infty} \frac{1}{n^s} = \prod_{p \in \mathcal{P}} (1 - p^{-s})^{-1}. \tag{1.6}$$

The function

$$\zeta(s) := \sum_{n=1}^{+\infty} \frac{1}{n^s}$$

is called the *Riemann zeta function* and it plays a fundamental role in the study of prime numbers (see [102] or the short introductions in [8] or [10]).

Let us now see a different proof of (1.4), which we owe to Clarkson [56].

Second proof of Theorem 1.11   Let $p_1 < p_2 < p_3 < \ldots$ be all the prime numbers, and let us assume that $\sum_{p \in \mathcal{P}}^{+\infty} 1/p$ converges. Then there exists a positive integer k such that

$$\sum_{n>k} \frac{1}{p_n} < \frac{1}{2} . \tag{1.7}$$

Let $Q = p_1 p_2 \cdots p_k$. For no $\ell \in \mathbb{N}$ does there exist p_j (with $1 \leq j \leq k$) such that $p_j \mid (1 + \ell Q)$. Then the prime divisors of $1 + \ell Q$ must be found among $p_{k+1}, p_{k+2}, \ldots$ and, for every $N \geq 1$, we have

$$\sum_{\ell=1}^{N} \frac{1}{1 + \ell Q} \leq \sum_{m=1}^{+\infty} \left(\sum_{n>k} \frac{1}{p_n} \right)^m . \tag{1.8}$$

In order to prove (1.8), we start by showing that every term $(1 + \ell Q)^{-1}$ in the LHS appears also in the RHS. Let

$$\frac{1}{1 + \ell Q} \underset{\text{say}}{=} \frac{1}{p_{k+2}^3 \, p_{k+5} \, p_{k+9}^4} .$$

Then $\left(p_{k+2}^3 \, p_{k+5} \, p_{k+9}^4 \right)^{-1}$ appears inside $\left(\sum_{n>k} \frac{1}{p_n} \right)^8$. In order to end the proof of (1.8) we observe that the terms $1 + \ell Q$ are distinct. Then (1.7) implies

$$\sum_{\ell=1}^{N} \frac{1}{1 + \ell Q} \leq \sum_{m=1}^{+\infty} \left(\frac{1}{2} \right)^m = 1 .$$

This is impossible because $\sum_{\ell=1}^{+\infty} \frac{1}{1+\ell Q} = +\infty$. $\qquad\qquad\qquad\square$

Remark 1.12   We now want to estimate the divergence of the series $\sum_{p \in \mathcal{P}}^{+\infty} 1/p$. We start by proving the inequality

$$\prod_{\mathcal{P} \ni p \leq x} \left(1 - \frac{1}{p} \right)^{-1} \leq \prod_{\mathcal{P} \ni p \leq x} e^{p^{-1} + p^{-2}} . \tag{1.9}$$

Indeed, let $f(t) = (1 - t) e^{t + t^2}$. Then $f'(t) = t(1 - 2t) e^{t + t^2} \geq 0$ for every $t \in [0, 1/2]$. Since $f(0) = 1$, we have $f(t) \geq 1$, that is to say $\frac{1}{1-t} \leq e^{t+t^2}$ for every $t \in [0, 1/2]$. This implies (1.9). If we take logarithms on both sides, while recalling (1.5) and the inequalities[3]

$$\sum_{k \leq x} \frac{1}{k} = \sum_{k=1}^{[x]} \frac{1}{k} \geq \int_1^{[x]+1} \frac{1}{x} \, dx = \log\left([x] + 1 \right) > \log x ,$$

---

[3] The integral part $[\alpha]$ of a real number α is the largest integer smaller than or equal to α, for example, $[5] = 5$, $[e] = 2$, $[-\pi] = -4$.

we obtain

$$\log\left(\log x\right) < \log\left(\prod_{\mathcal{P} \ni p \le x} \left(1 - \frac{1}{p}\right)^{-1}\right) \le \log\left(\prod_{\mathcal{P} \ni p \le x} e^{p^{-1}+p^{-2}}\right)$$
$$= \sum_{\mathcal{P} \ni p \le x} \left(p^{-1} + p^{-2}\right) \, .$$

Since

$$\sum_{\mathcal{P} \ni p \le x} p^{-2} < \sum_{2 \le k \le x} k^{-2} < \int_1^x t^{-2} dt < 1 \, ,$$

we obtain

$$\sum_{\mathcal{P} \ni p \le x} \frac{1}{p} > \log\left(\log x\right) - 1 \, .$$

Remark 1.13 A similar inequality holds from above. More precisely, Mertens has proved in 1874 that

$$\sum_{\mathcal{P} \ni p \le x} \frac{1}{p} = \log\left(\log x\right) + M + O\left(\frac{1}{\log x}\right) \, ,$$

where $M = 0.26149\cdots$

1.2 Chebyshev's theorem

For every real number $x \ge 2$ let[4]

$$\pi(x) := \sum_{\mathcal{P} \ni p \le x} 1 = \operatorname{card}\left(\mathcal{P} \cap [2, x]\right) \, .$$

The celebrated *prime number theorem* (independently proved by Hadamard and de la Vallée Poussin in 1896) says that

$$\pi(x) \sim \frac{x}{\log x}$$

$(a(x) \sim b(x)$ means that $\frac{a(x)}{b(x)} \to 1$) as $x \to +\infty$. This theorem had already been conjectured by Gauss and Legendre at the beginning of the nineteenth century. See, for example, [8] and [128] for an analytic and an 'elementary' proof, respectively. See [128, 9.5] for some historical remarks on the elementary proof. See also [83] for a brief introduction to the history of the prime number theorem.

---

[4] If A is a finite set, then card(A) is the number of elements in A.

Around 1850 Chebyshev proved a weaker result, i.e. the existence of two positive constants c_1 and c_2 such that

$$c_1 \frac{x}{\log x} \le \pi(x) \le c_2 \frac{x}{\log x} . \tag{1.10}$$

Before proving (1.10) we need a lemma.

Lemma 1.14 *Let n be a positive integer and let $p \in \mathcal{P}$. Then the number*

$$\sum_{j \ge 1} \left[\frac{n}{p^j} \right] \tag{1.11}$$

is the exponent (possibly zero) of p in the canonical decomposition of $n!$.

Observe that the sum in (1.11) is always a finite sum.[5]

Proof If $p > n$, then p does not appear in the canonical decomposition of $n!$ (and $\left[n/p^j \right] = 0$ for every $j \ge 1$). If $p \le n$, then the $[n/p]$ numbers

$$p, 2p, 3p, \ldots, \left[\frac{n}{p} \right] p \tag{1.12}$$

are all the positive integers not larger than n and that can be divided by p. Observe that $\left[n/p^2 \right]$ numbers among those in (1.12) can also be divided by p^2. They are

$$p^2, 2p^2, 3p^2, \ldots, \left[\frac{n}{p^2} \right] p^2$$

(of course $\left[n/p^2 \right] = 0$ if $p^2 > n$). Then the power of p in the canonical decomposition of $n!$ is at least $[n/p] + \left[n/p^2 \right]$. Going on, we observe that the following $\left[n/p^3 \right]$ numbers can be divided by p^3:

$$p^3, 2p^3, 3p^3, \ldots, \left[\frac{n}{p^3} \right] p^3 .$$

Then the power of p in the canonical decomposition of $n!$ is at least $[n/p] + \left[n/p^2 \right] + \left[n/p^3 \right]$. In this way we see that the finite sum $\sum_{j \ge 1} \left[n/p^j \right]$ is the exponent of p in the canonical decomposition of $n!$ □

---

[5] The above lemma can be used to compute the canonical factorization of $n!$. Consider, for example, 10! The prime numbers not exceeding 10 are $2, 3, 5, 7$. We have

$$\left[\frac{10}{2} \right] + \left[\frac{10}{2^2} \right] + \left[\frac{10}{2^3} \right] = 8 , \quad \left[\frac{10}{3} \right] + \left[\frac{10}{3^2} \right] = 4 , \quad \left[\frac{10}{5} \right] = 2 , \quad \left[\frac{10}{7} \right] = 1 .$$

Then

$$10! = 2^8 \cdot 3^4 \cdot 5^2 \cdot 7 .$$

Theorem 1.15 (Chebyshev) *There exist two positive constants c_1, c_2 such that, for every real number $x \geq 2$,*

$$c_1 \frac{x}{\log x} \leq \pi(x) \leq c_2 \frac{x}{\log x} . \tag{1.13}$$

Proof The main idea is to study the difference

$$\pi(2n) - \pi(n) = \operatorname{card}(\mathcal{P} \cap (n, 2n])$$

for every integer $n \geq 2$. For this purpose we use the binomial coefficient

$$\binom{2n}{n} = \frac{2n(2n-1)\cdots(n+1)}{n(n-1)\cdots 1} \tag{1.14}$$

(i.e., the 'bisector' of Pascal's (or Tartaglia's) triangle) and observe that every number $p \in \mathcal{P} \cap (n, 2n]$ divides the numerator in (1.14), but does not divide the denominator. Then $p \mid \binom{2n}{n}$, hence also $P_n := \prod_{p \in \mathcal{P} \cap (n, 2n]} p$ divides $\binom{2n}{n}$. Then we have

$$n^{\pi(2n) - \pi(n)} < P_n \leq \binom{2n}{n} . \tag{1.15}$$

For every prime number p let r_p be the integer satisfying

$$p^{r_p} \leq 2n < p^{r_p + 1} . \tag{1.16}$$

We use Lemma 1.14 to compute the power of p in the canonical decomposition of $\binom{2n}{n}$. Indeed, we know that p appears with exponent $\sum_{j=1}^{+\infty} \left[2n/p^j\right]$ in $(2n)!$ and with exponent $2\sum_{j=1}^{+\infty} \left[n/p^j\right]$ in $(n!)^2$. Then it appears with exponent

$$\sum_{j=1}^{+\infty} \left(\left[\frac{2n}{p^j}\right] - 2\left[\frac{n}{p^j}\right] \right)$$

in the canonical decomposition of $\binom{2n}{n} = \frac{(2n)!}{(n!)^2}$. Now observe that, for every real number α, the function $\alpha \mapsto [\alpha] - 2\left[\frac{\alpha}{2}\right]$ takes only the values 0 or 1. Indeed, $[\alpha] - 2\left[\frac{\alpha}{2}\right] \in \mathbb{Z}$ and

$$-1 = \alpha - 1 - 2\frac{\alpha}{2} < [\alpha] - 2\left[\frac{\alpha}{2}\right] < \alpha - 2\left(\frac{\alpha}{2} - 1\right) = 2 .$$

Then (1.16) implies

$$0 \leq \sum_{j=1}^{+\infty} \left(\left[\frac{2n}{p^j}\right] - 2\left[\frac{n}{p^j}\right] \right) \leq \sum_{\{j : p^j \leq 2n\}} 1 = r_p .$$

Hence

$$\binom{2n}{n} = \prod_{p \in \mathcal{P} \cap [2, 2n]} p^{\sum_{j=1}^{+\infty} \left(\left[\frac{2n}{p^j}\right] - 2\left[\frac{n}{p^j}\right] \right)}$$

divides

$$\prod_{p \in \mathcal{P} \cap [2,2n]} p^{r_p} \, . \tag{1.17}$$

The last product has $\pi(2n)$ terms. Then (1.16) gives

$$\binom{2n}{n} \leq \prod_{p \in \mathcal{P} \cap [2,2n]} p^{r_p} \leq (2n)^{\pi(2n)} \, . \tag{1.18}$$

Let us now estimate[6] $\binom{2n}{n}$. Observe that

$$\frac{\binom{m}{k}}{\binom{m}{k-1}} = \frac{\frac{m!}{k!(m-k)!}}{\frac{m!}{(k-1)!(m-k+1)!}} = \frac{m-k+1}{k} > 1 \iff k < \frac{m+1}{2} \, .$$

Then $\binom{2n}{n}$ is the largest binomial coefficient in the expansion of $(1+1)^{2n}$. Hence

$$2^{2n} = (1+1)^{2n} = \sum_{k=0}^{2n} \binom{2n}{k} = 2 + \sum_{k=1}^{2n-1} \binom{2n}{k} \leq 2 + (2n-1)\binom{2n}{n} \leq 2n \binom{2n}{n} \, .$$

Then

$$\binom{2n}{n} \geq \frac{2^{2n}}{2n} \, . \tag{1.19}$$

Moreover, we obviously have

$$\binom{2n}{n} < \sum_{k=0}^{2n} \binom{2n}{k} = 2^{2n} \tag{1.20}$$

(this upper bound will be used later). Then (1.18) and (1.19) give

$$2^n \leq \frac{2^{2n}}{2n} \leq (2n)^{\pi(2n)} \, ,$$

that is,

$$\pi(2n) \geq \frac{n \log 2}{\log(2n)} \, .$$

From this we deduce that for every real number $x \geq 2$ we have

$$\pi(x) \geq \frac{1}{4} \log 2 \, \frac{x}{\log x} \, . \tag{1.21}$$

---

[6] Observe that a simple use of Stirling's formula ($n! \sim \sqrt{2\pi n}\,(n/e)^n$, see Theorem 6.32) gives

$$\binom{2n}{n} \sim 4^n \, (n\pi)^{-1/2} \, .$$

As for the estimate from above, we observe that (1.15) and (1.20) imply

$$n^{\pi(2n)-\pi(n)} < 2^{2n} \, ,$$

that is,

$$(\pi(2n) - \pi(n)) \log n < 2n \log 2 \tag{1.22}$$

or

$$\pi(2n) < \frac{2n \log 2}{\log n} + \pi(n) \, . \tag{1.23}$$

We are going to show that for every integer $k \geq 2$ we have

$$\pi(k) \leq 16 \log 2 \, \frac{k}{\log k} \, . \tag{1.24}$$

Indeed, (1.24) is true for $k = 2$ and the function $x \mapsto \frac{x}{\log x}$ increases when $x \geq e$. Then

$$\pi(k) \leq \pi(16) = 6 < 16 \log 2 \, \frac{3}{\log 3} \leq 16 \log 2 \, \frac{k}{\log k}$$

for every $3 \leq k \leq 16$. Now let $k > 16$. We proceed by induction assuming (1.24) true for all indices between 2 and $k - 1$. Let n be the positive integer satisfying $2n - 2 < k \leq 2n$. Then (1.23) implies

$$\pi(k) \leq \pi(2n) < \frac{2n \log 2}{\log n} + \pi(n) \, .$$

Since $n < k$, the induction hypothesis gives

$$\pi(k) < \frac{2n \log 2}{\log n} + 16 \log 2 \, \frac{n}{\log n} = 18 \log 2 \, \frac{n}{\log n} \, .$$

Since $k > 16$, we have

$$k^{3/4} < \frac{k}{2} < n < \frac{k+2}{2} < \frac{9}{16} k \, .$$

Then

$$\pi(k) < 18 \log 2 \, \frac{\frac{9}{16} k}{\frac{3}{4} \log k} < 16 \log 2 \, \frac{k}{\log k} \, .$$

This readily gives

$$\pi(x) \leq 32 \log 2 \, \frac{x}{\log x} \, .$$

$\square$

Another proof of the estimate from below in (1.13)   For every positive integer
n, let Q_n be the least common multiple of the numbers

$$n + 1, n + 2, \ldots, 2n + 1 .$$

Since $0 \le x(1 - x) \le 1/4$ for every $0 \le x \le 1$, we have

$$\frac{1}{4^n} > \int_0^1 \left(x - x^2\right)^n dx = \int_0^1 \sum_{j=0}^n (-1)^j \binom{n}{j} x^{n+j} \, dx$$

$$= \sum_{j=0}^n (-1)^j \binom{n}{j} \frac{1}{n + j + 1}$$

$$= \frac{1}{n + 1} - n \frac{1}{n + 2} + \frac{n(n + 1)}{2} \frac{1}{n + 3} - \ldots + (-1)^n \frac{1}{2n + 1}$$

$$:= \frac{P_n}{Q_n} \ge \frac{1}{Q_n}$$

(observe that P_n is a positive integer). Then $Q_n > 4^n$. We write the canonical
decomposition of Q_n :

$$Q_n = p_1^{m_1} p_2^{m_2} \cdots p_N^{m_N} .$$

The choice of Q_n implies $p_j^{m_j} \le 2n + 1$ for every j, then $N \le \pi(2n + 1)$ and

$$4^n < Q_n \le (2n + 1)^{\pi(2n+1)} ,$$

$$\pi(2n + 1) > 2 \log 2 \, \frac{n}{\log(2n + 1)} .$$

This implies

$$\pi(x) \ge c_1 \frac{x}{\log x}$$

for every $x \ge 2$.

This proof is due to Tihomirov.   □

Remark 1.16   In (1.22) we have seen that

$$\pi(2n) - \pi(n) < \log 4 \, \frac{n}{\log n}$$

for every integer $n \ge 2$. It is possible to prove a similar bound from below:

$$\pi(2n) - \pi(n) > c \, \frac{n}{\log n} .$$

This last inequality is a strong form of the so-called *Bertrand's postulate* (per-
haps more correctly called the *Bertrand–Chebyshev theorem*, see [143]), which
states that for every positive integer n there is always at least one prime number
p such that $n < p \le 2n$. Bertrand's postulate is not a trivial result, since there

exist arbitrarily long sequences of consecutive composite numbers (for every large integer k, consider $k! + 2, k! + 3, \ldots, k! + k$).

Zhang has recently proved (see [183]) the following fundamental result.

Theorem 1.17 (Zhang) *Let p_n be the nth prime number. Then*

$$\liminf_{n \to +\infty} (p_{n+1} - p_n) < +\infty \, .$$

Exercises

1) Let $1, 1, 2, 3, 5, 8, 13, 21, \ldots$ be the Fibonacci sequence, defined by $a_1 = a_2 = 1$ and $a_n = a_{n-1} + a_{n-2}$ if $n \geq 3$. Prove that $(a_{n+1}, a_n) = 1$ for every n.

2) Find the positive integers M which have 6 as their first digit and are equal to $M/25$ once the first digit is erased.

3) Let $a, b, c \in \mathbb{Z}$. Show that the equation $ax + by = c$ has an integer solution if and only if $(a, b) \mid c$.

4) Prove that if $2^n + 1$ is a prime number, then n is a power of 2.

5) Prove that for all positive integers a and b we have

$$\left(2^a - 1, 2^b - 1\right) = 2^{(a,b)} - 1 \, .$$

6) Prove that 10 is the only positive composite number such that each one of its non-trivial divisors has the form $a^r + 1$ with integers $a \geq 1$ and $r \geq 2$.

7) Let a, b be two positive integers with $(a, b) = 1$. Determine the largest integer d that cannot be written as $d = ma + nb$ with m and n positive integers. This is part of the so-called *coin problem*. (Why this name?)

8) Let $k \geq 1$ and let $a_0, a_1, \ldots, a_k \in \mathbb{Z}$, with $a_k \neq 0$. For every positive integer n, let $Q(n) = \sum_{j=0}^{k} a_j n^j$. Prove that there exist infinitely many integers n_0 such that $Q(n_0) \notin \mathcal{P}$.

9) Prove (1.6).

10) Prove that

$$\zeta(s) \sim \frac{1}{s-1}$$

as $s \to 1^+$.

11) Euclid's proof of the infinitude of prime numbers shows that the (increasing) sequence p_n of prime numbers satisfies

$$p_{n+1} \leq 1 + \prod_{j=1}^{N} p_j \, .$$

Deduce the following result:

$$\pi(x) \geq c \, \log(\log x) \ .$$

12) Write the prime numbers as an increasing sequence $p_1 < p_2 < \ldots$ and prove the existence of two positive constants c_1 and c_2 such that

$$c_1 \, n \log(n) \leq p_n \leq c_2 \, n \log(n) \ .$$

2

Arithmetic functions and integer points

In number theory, an *arithmetic function* is a real or complex-valued function $f(n)$ defined on $\mathbb{N}$. Many of these functions represent important arithmetic properties of n; in particular, we shall see the Dirichlet function $d(n)$, which counts the positive divisors of n, the function $r(n)$, which gives the number of representations of n as a sum of two squares, and the Euler function $\phi(n)$, which says how many numbers between 1 and n are coprime with n.

Several important arithmetic functions show an irregular behaviour as n grows. Consider, for example, $d(n)$: if n is a prime number, then $d(n) = 2$, but this does not exclude that the number $n + 1$ may have many divisors, so that $d(n + 1)$ is large. It may therefore be difficult to consider the behaviour of a given arithmetic function $f(n)$ as n tends to infinity. Sometimes it is useful to study the average

$$n \longmapsto \frac{1}{n} \sum_{j=1}^{n} f(j) \, ,$$

as $n \to +\infty$. In certain cases the study of this arithmetic mean consists in counting the *integer points* (that is, the points with integral coordinates) inside suitable domains contained in the plane or in $\mathbb{R}^d$. As a simple yet amazing result, we shall see that (on average) a natural number n can be written as a sum of two squares in ... π ways.

Several important arithmetic functions have algebraic properties, which we are going to introduce in a general setting.

2.1 Arithmetic functions and Dirichlet product

We define the sum of two arithmetic functions in the obvious way:

$$(f + g)(n) := f(n) + g(n)$$

19

(pointwise sum). As for the product, besides the pointwise product

$$(fg)(n) := f(n)g(n) \, ,$$

we consider the *Dirichlet convolution* (or *Dirichlet product*)

$$(f * g)(n) := \sum_{d|n} f(d)g(n/d) = \sum_{dd'=n} f(d)g(d') \, , \qquad (2.1)$$

where the sums $\sum_{d|n}$ and $\sum_{dd'=n}$ are over the positive divisors d of n.

Here we can see four simple arithmetic functions

$$I(n) := \begin{cases} 1 & \text{if } n = 1, \\ 0 & \text{if } n > 1, \end{cases}$$

$$0(n) := 0 \ \text{ for every } n \, , \qquad (2.2)$$

$$u(n) := 1 \ \text{ for every } n \, ,$$

$$N(n) := n.$$

Theorem 2.1 *The set of all arithmetic functions, with pointwise sum and Dirichlet product, is a commutative ring with additive identity* $0(n)$ *and multiplicative identity* $I(n)$.

Proof It is obvious that the set of all arithmetic functions, with the above sum, is an abelian group with identity $0(n)$. We need to show that

(i) $f * g = g * f$,
(ii) $(f * g) * h = f * (g * h)$,
(iii) $f * (g + h) = f * g + f * h$,
(iv) $f * I = f$.

The identity (i) follows from the symmetry of the second sum in (2.1). In order to prove (ii) we write

$$((f * g) * h)(n) = \sum_{dd'=n} (f * g)(d)h(d') = \sum_{dd'=n} \left(\sum_{pq=d} f(p)g(q) \right) h(d')$$

$$= \sum_{pqd'=n} f(p)g(q)h(d')$$

and observe the symmetry of p, q, d' in the last sum. We finally prove (iii)

$$(f * (g + h))(n) = \sum_{d|n} f(d)g(n/d) + \sum_{d|n} f(d)h(n/d)$$

$$= (f * g)(n) + (f * h)(n)$$

and (iv)

$$(f * I)(n) = \sum_{d|n} f(d) I(n/d) = f(n) I(1) = f(n) .$$

□

We now introduce the *Dirichlet inverse* of an arithmetic function.

Theorem 2.2 *Let f be an arithmetic function such that $f(1) \neq 0$. Then there exists one and only one arithmetic function f^{*-1} (we call it the Dirichlet inverse of f) such that*

$$f * f^{*-1} = I .$$

Proof We shall define $f^{*-1}(n)$ by induction on n. For $n = 1$ let $f^{*-1}(1) = 1/f(1)$. Now assume that $f^{*-1}(k)$ has been introduced for every $1 \leq k < n$. Observe that f^{*-1} must satisfy

$$\sum_{d|n} f(n/d) f^{*-1}(d) = 0$$

(for $n > 1$). That is to say

$$0 = f(1) f^{*-1}(n) + \sum_{\substack{d|n \\ d<n}} f(n/d) f^{*-1}(d)$$

or

$$f^{*-1}(n) = -\frac{1}{f(1)} \sum_{\substack{d|n \\ d<n}} f(n/d) f^{*-1}(d) . \tag{2.3}$$

Then f^{*-1} is defined. □

Observe that if $p \in \mathcal{P}$, then (2.3) implies the following simple identity:

$$f^{*-1}(p) = -\frac{1}{f^2(1)} f(p) .$$

The following property is satisfied by several important arithmetic functions.

Definition 2.3 An arithmetic function f is called a *multiplicative function* if it is not identically zero and

$$f(mn) = f(m) f(n) \tag{2.4}$$

whenever $(m, n) = 1$.

Remark 2.4 If f is a multiplicative function, then $f(1) = 1$. Indeed, (2.4) implies $f(1) f(1) = f(1)$, hence $f(1) = 0$ or $f(1) = 1$. Observe that $f(1) = 0$ implies $f(n) = f(1) f(n) = 0$ for every n, against the assumption on f.

If f is a multiplicative function and n has canonical decomposition $n = p_1^{m_1} p_2^{m_2} \cdots p_N^{m_N}$, then

$$f(n) = \prod_{j=1}^{N} f(p^{m_j}),$$

so that it may be enough to compute $f(p^s)$ for $p \in \mathcal{P}$ and s a positive integer.

Theorem 2.5   *Let f and g be arithmetic functions.*

   (i) *If f and g are multiplicative functions, then $h = f * g$ is also multiplicative.*
   (ii) *If f and $f * g$ are multiplicative functions, then g is also multiplicative.*
   (iii) *If f is a multiplicative function, then f^{*-1} is also multiplicative.*

Proof (i) If $(m, n) = 1$ and $d \mid mn$, then d can be written in a unique way as $d = d_1 d_2$, where the natural numbers d_1 and d_2 satisfy $d_1 \mid m$, $d_2 \mid n$ (and of course $(d_1, d_2) = 1$). Then the assumptions on f and g imply

$$h(mn) = \sum_{d \mid mn} f(d) g(mn/d) = \sum_{d_1 \mid m,\ d_2 \mid n} f(d_1 d_2) g((m/d_1)(n/d_2))$$

$$= \sum_{d_1 \mid m} \sum_{d_2 \mid n} f(d_1) f(d_2) g(m/d_1) g(n/d_2)$$

$$= \sum_{d_1 \mid m} f(d_1) g(m/d_1) \sum_{d_2 \mid n} f(d_2) g(n/d_2) = h(m) h(n) .$$

(ii)   Assume that g is not a multiplicative function. Then there exist two co-prime natural numbers m_0 and n_0 such that $g(m_0 n_0) \neq g(m_0) g(n_0)$. Let us choose m_0 and n_0 so that their product $m_0 n_0$ is as small as possible. Observe that $m_0 n_0 > 1$, otherwise we should have $g(1) \neq (g(1))^2$, which is impossible because Remark 2.4 implies

$$1 = (f * g)(1) = f(1) g(1) = g(1) .$$

Then $m_0 n_0 > 1$ and $g(ab) = g(a) g(b)$ for every pair of positive integers a, b such that $(a, b) = 1$ and $ab < m_0 n_0$. Let d be a positive divisor of $m_0 n_0$. Again we write $d = d_1 d_2$, with $d_1 \mid m_0$, $d_2 \mid n_0$. Then

$$(f * g)(m_0 n_0) = \sum_{d \mid m_0 n_0} f(d) g(m_0 n_0/d)$$

$$= f(1) g(m_0 n_0) + \sum_{d_1 \mid m_0,\ d_2 \mid n_0,\ d_1 d_2 > 1} f(d_1 d_2) g(m_0 n_0/d_1 d_2)$$

$$= g(m_0 n_0) - g(m_0) g(n_0) + \sum_{d_1 \mid m_0,\ d_2 \mid n_0} f(d_1) f(d_2) g((m_0/d_1)(n_0/d_2))$$

$$= g(m_0 n_0) - g(m_0) g(n_0) + \sum_{d_1 | m_0} f(d_1) g(m_0/d_1) \sum_{d_2 | n_0} f(d_2) g(n_0/d_2)$$

$$= g(m_0 n_0) - g(m_0) g(n_0) + (f * g)(m_0)(f * g)(n_0) .$$

Since $g(m_0 n_0) \neq g(m_0) g(n_0)$ we deduce that

$$(f * g)(m_0 n_0) \neq (f * g)(m_0)(f * g)(n_0) .$$

This is impossible if $f * g$ is a multiplicative function.

(iii) f and $I = f * f^{*-1}$ are multiplicative functions. Therefore (iii) follows from (ii). □

Corollary 2.6 *Let f be an arithmetic function. Then*

$$k(n) = \sum_{m | n} f(m)$$

is a multiplicative function if and only if the same is true for f.

Proof We may write $k = f * u$ (see (2.2)). Since u is a multiplicative function, Theorem 2.5 says that k is a multiplicative function if and only if the same is true for f. □

2.2 $\mu(n)$

We now introduce the *Möbius function* $\mu(n)$:

$$\mu(n) := \begin{cases} 1 & \text{if } n = 1, \\ (-1)^N & \text{if } n > 1 \text{ is a product of } N \text{ distinct prime numbers,} \\ 0 & \text{otherwise.} \end{cases}$$

Then $\mu(1) = 1$, $\mu(2) = -1$, $\mu(3) = -1$, $\mu(4) = 0$, $\mu(5) = -1$, $\mu(6) = 1$, $\mu(7) = -1$, $\mu(8) = 0$, ... , $\mu(12) = 0$, ... , $\mu(30) = -1$, ...

The arithmetic function μ is multiplicative. Indeed, assume that $(m, n) = 1$. Then mn admits a prime number p such that $p^2 | mn$ if and only if at least one number between m and n has the same property. That is to say, $\mu(mn) = 0$ if and only if $\mu(m)\mu(n) = 0$. Now let m be the product of M distinct prime numbers and let n be the product of N distinct prime numbers. Then (since $(m, n) = 1$) the number mn is the product of $M + N$ distinct prime numbers, and

$$\mu(mn) = (-1)^{M+N} = (-1)^M (-1)^N = \mu(m)\mu(n) .$$

We are going to show that u is the Dirichlet inverse of μ.

Lemma 2.7 $\mu^{*-1} = u$, *that is,*

$$\sum_{m|n} \mu(m) = \begin{cases} 1 & \text{if } n = 1, \\ 0 & \text{if } n > 1. \end{cases} \tag{2.5}$$

Proof The identity (2.5) is obvious if $n = 1$. By Corollary 2.6, the function $n \mapsto \sum_{m|n} \mu(m)$ is a multiplicative function. Then it is enough to prove (2.5) when n is a power of a prime number, say $n = p^s$ ($s \geq 1$). Indeed, in this case

$$\sum_{m|p^s} \mu(m) = \sum_{j=0}^{s} \mu\left(p^j\right) = 1 - 1 + 0 + 0 + \cdots + 0 = 0 \,.$$

$\square$

We introduce the *Möbius inversion formula*, which was proved by Möbius in 1832.

Theorem 2.8 (Möbius' inversion formula) *Let f and g be arithmetic functions. Then $g = u * f$ if and only if $f = \mu * g$.*

Proof Lemma 2.7 implies

$$\mu * g = \mu * f * u = f * (\mu * u) = f * I = f \,.$$

The converse is similar. $\square$

We are going to apply the Möbius function to the study of square-free numbers. A natural number is a *square-free number* if it is not divisible by the square of any integer > 1. Let Q denote the set of square-free numbers. For every positive real number x let

$$Q(x) := \text{card}\,\{n \in Q : n \leq x\} \,.$$

The following result was proved by Gegenbauer in 1885.

Theorem 2.9 (Gegenbauer) *As $x \to +\infty$ we have*

$$Q(x) = \frac{6}{\pi^2} x + O\left(\sqrt{x}\right) \,. \tag{2.6}$$

Proof Observe that

$$\mu^2(n) = \begin{cases} 1 & \text{if } n \text{ is square-free,} \\ 0 & \text{otherwise.} \end{cases}$$

Then

$$Q(x) = \sum_{n \leq x} \mu^2(n) \,.$$

Let $g = \mu^2 * \mu$. Then the Möbius inversion formula implies $\mu^2 = u * g$. By

Theorem 2.5, g is a multiplicative function. Let p be a prime number and let $\ell \geq 1$, then

$$g\left(p^{\ell}\right) = \sum_{k=0}^{\ell} \mu^2\left(p^k\right)\mu\left(p^{\ell-k}\right) = \begin{cases} -1 & \text{if } \ell = 2, \\ 0 & \text{otherwise.} \end{cases}$$

Then $g(n) = 0$ unless $n = m^2$, where m is square-free. Hence $g\left(m^2\right) = \mu(m)$ for every $m \in \mathbb{N}$. Then

$$\sum_{n\leq x}\mu^2(n) = \sum_{n\leq x}(u * g)(n) = \sum_{n\leq x}\sum_{d|n}u\left(\frac{n}{d}\right)g(d) = \sum_{d\leq x}g(d)\sum_{\substack{n\leq x \\ d|n}}1$$

$$= \sum_{d\leq x}g(d)\left[\frac{x}{d}\right] = \sum_{m\leq \sqrt{x}}\mu(m)\left(\frac{x}{m^2} + O(1)\right)$$

$$= x\sum_{m=1}^{+\infty}\frac{\mu(m)}{m^2} - x\sum_{m>\sqrt{x}}\frac{\mu(m)}{m^2} + O\left(\sqrt{x}\right) = x\sum_{m=1}^{+\infty}\frac{\mu(m)}{m^2} + O\left(\sqrt{x}\right).$$

We claim that

$$\sum_{m=1}^{+\infty}\frac{\mu(m)}{m^2} = \frac{6}{\pi^2}. \tag{2.7}$$

Indeed, by (2.5) we have

$$1 = \sum_{k=1}^{+\infty}\frac{1}{k^2}\left(\sum_{m|k}\mu(m)\right) = \sum_{k=1}^{+\infty}\frac{1}{k^2}\left(\sum_{mn=k}\mu(m)\right) = \sum_{m,n=1}^{+\infty}\frac{\mu(m)}{(mn)^2}$$

$$= \sum_{n=1}^{+\infty}\frac{1}{n^2}\sum_{m=1}^{+\infty}\frac{\mu(m)}{m^2} = \frac{\pi^2}{6}\sum_{m=1}^{+\infty}\frac{\mu(m)}{m^2},$$

by (4.35). This implies (2.6). □

Since

$$\frac{Q(x)}{x} = \frac{6}{\pi^2} + O\left(x^{-1/2}\right)$$

we may say that the percentage of square-free numbers is $6/\pi^2$.

2.3 $d(n)$ and $\sigma(n)$

The *Dirichlet function* $d(n)$ counts the positive divisors of n :

$$d(n) := \sum_{m|n}1. \tag{2.8}$$

Then $d(1) = 1$, $d(2) = 2$, $d(3) = 2$, $d(4) = 3$, $d(5) = 2$, $d(6) = 4$, ...

On the one hand, we have $d(p) = 2$ if and only if p is a prime number. On the other hand, we have $d(q) = 2^N$ if q is a product of N distinct prime numbers.

We also introduce the arithmetic function

$$\sigma(n) := \sum_{m|n} m , \qquad (2.9)$$

which sums the positive divisors of n. Then $\sigma(1) = 1$, $\sigma(2) = 3$, $\sigma(3) = 4$, $\sigma(4) = 7$, $\sigma(5) = 6$, $\sigma(6) = 12$, ... Observe that $\sigma(p) = p + 1$ if and only if p is a prime number.

Proposition 2.10 *Write* $n = p_1^{m_1} p_2^{m_2} \cdots p_N^{m_N}$ *according to its canonical decomposition. Then*

$$d(n) = (m_1 + 1)(m_2 + 1) \cdots (m_N + 1) \qquad (2.10)$$

and

$$\sigma(n) = \frac{p_1^{m_1+1} - 1}{p_1 - 1} \frac{p_2^{m_2+1} - 1}{p_2 - 1} \cdots \frac{p_N^{m_N+1} - 1}{p_N - 1} .$$

Hence $d(n)$ *and* $\sigma(n)$ *are multiplicative functions.*

Proof If $k \mid n$, then $k = p_1^{\ell_1} p_2^{\ell_2} \cdots p_N^{\ell_N}$ with $0 \leq \ell_j \leq m_j$ for every j. Then

$$d(n) = \operatorname{card}\left\{(\ell_1, \ell_2, \ldots, \ell_N) : 0 \leq \ell_j \leq m_j, \ j = 1, \ldots m_j\right\}$$
$$= (m_1 + 1)(m_2 + 1) \cdots (m_N + 1) .$$

In a similar way,

$$\sigma(n) = \sum_{\ell_1=0}^{m_1} \sum_{\ell_2=0}^{m_2} \cdots \sum_{\ell_N=0}^{m_N} p_1^{\ell_1} p_2^{\ell_2} \cdots p_N^{\ell_N} = \left(\sum_{\ell_1=0}^{m_1} p_1^{\ell_1}\right)\left(\sum_{\ell_2=0}^{m_2} p_2^{\ell_2}\right) \cdots \left(\sum_{\ell_N=0}^{m_N} p_N^{\ell_N}\right)$$
$$= \frac{p_1^{m_1+1} - 1}{p_1 - 1} \frac{p_2^{m_2+1} - 1}{p_2 - 1} \cdots \frac{p_N^{m_N+1} - 1}{p_N - 1} .$$

$\square$

The natural numbers n which satisfy $\sigma(n) = 2n$ are called *perfect numbers*. $6 = 1 + 2 + 3$ and $28 = 1 + 2 + 4 + 7 + 14$ are perfect numbers. The existence of at least one odd perfect number is an open problem. For even perfect numbers we have the following result, due to Euclid and Euler.

Theorem 2.11 (Euclid–Euler) *An even integer* n *is a perfect number if and only if it can be written in the form*

$$n = 2^{p-1}(2^p - 1) ,$$

where $2^p - 1$ *is a prime number.*[1]

Proof Let $n = 2^{p-1}(2^p - 1)$ and assume that $2^p - 1$ is a prime number. Since σ is a multiplicative function and $\left(2^{p-1}, 2^p - 1\right) = 1$, we have

$$\sigma(n) = \sigma\left(2^{p-1}\right)\sigma(2^p - 1) = (1 + 2 + 2^2 + \ldots + 2^{p-1})\,\sigma(2^p - 1)$$
$$= (2^p - 1)\,2^p = 2n \;.$$

Hence n is perfect. Conversely, let n be an even perfect number. We can write $n = 2^{p-1}q$, with $p > 1$ and q odd. Then

$$2^p q = 2n = \sigma(n) = \sigma\left(2^{p-1}\right)\sigma(q) = (2^p - 1)\sigma(q) \;.$$

Hence $(2^p - 1) \mid q$, and we write $q = (2^p - 1)\,s$. Then

$$\sigma(q) = 2^p s \;.$$

Now observe that s and $q = (2^p - 1)\,s$ are two different divisors of q, and

$$q + s = 2^p s = \sigma(q) \;.$$

Then q and s are the only divisors of q. Hence q is a prime number and $s = 1$. This gives $q = 2^p - 1$ and $n = (2^p - 1)\,2^{p-1}$, where $2^p - 1$ is a prime number. □

Remark 2.12 The previous result is very elegant, but it is not as useful as it may appear. The problem is that we do not know which numbers of the form $2^m - 1$ are prime numbers. We have seen that $2^m - 1$ is a prime number only if m is a prime number, but the converse is false: $2^5 - 1 = 31$ is a prime number, $2^7 - 1 = 127$ is a prime number, but $2^{11} - 1 = 2047 = 23 \cdot 89$ is not a prime number. Actually we do not know whether there are infinitely many prime numbers of this form. Therefore we do not know whether there are infinitely many even perfect numbers. These prime numbers are called Mersenne primes (after the French monk Mersenne, who investigated them in the seventeenth century). Up to the first half of the twentieth century only a few Mersenne primes were known. Then computers helped mathematicians (the first was Robinson in 1952 at UCLA) to find several new numbers, and at the moment we know 48 Mersenne primes. The largest one is $2^{57885161} - 1$, and it was discovered by Cooper in 2013 at the University of Central Missouri. Nowadays the search for

---

[1] Observe that if $2^p - 1$ is a prime number, then the same is true for p. Indeed, assume that $p = k\ell$ with $k > 1$ and $\ell > 1$. Then

$$2^p - 1 = \left(2^k\right)^\ell - 1 = (2^k - 1)\sum_{j=0}^{\ell-1} 2^{kj} \;.$$

Mersenne primes involves distributed computing and thousands of volunteers connected to the Great Internet Mersenne Prime Search (GIMPS). See [184].

The following results will provide some information on the behaviour of $d(n)$ and $\sigma(n)$ when n is large.

Theorem 2.13   *We have the following estimates.*

(i) *For every $\varepsilon > 0$ there exists a constant $c = c_\varepsilon$ such that*

$$d(n) \le c\, n^\varepsilon$$

for every $n \ge 1$.

(ii) *For no constants c_1 and c_2 can the inequality*

$$d(n) \le c_1 \log^{c_2} n$$

be true for every $n \ge 2$.

Proof   In order to prove (i) we may assume that $0 < \varepsilon < 1$. If n has canonical decomposition $n = p_1^{m_1} p_2^{m_2} \cdots p_N^{m_N}$, then Proposition 2.10 implies

$$\frac{d(n)}{n^\varepsilon} = \frac{1 + m_1}{p_1^{\varepsilon m_1}} \frac{1 + m_2}{p_2^{\varepsilon m_2}} \cdots \frac{1 + m_N}{p_N^{\varepsilon m_N}} .$$

Since $p_j \ge 2$ for every j, we have

$$p_j^{\varepsilon m_j} \ge 2^{\varepsilon m_j} = e^{\varepsilon m_j \log 2} \ge 1 + \varepsilon m_j \log 2 > \left(1 + m_j\right) \varepsilon \log 2 .$$

Hence

$$\frac{1 + m_j}{p_j^{\varepsilon m_j}} < \frac{1}{\varepsilon \log 2}$$

for every j. If we consider only the prime numbers p_j which satisfy $p_j \ge 2^{1/\varepsilon}$, then

$$\frac{1 + m_j}{p_j^{\varepsilon m_j}} \le \frac{1 + m_j}{2^{m_j}} \le 1$$

(because $1 + \ell \le 2^\ell$ for every positive integer ℓ). Then

$$\frac{d(n)}{n^\varepsilon} \le \prod_{p_j < 2^{1/\varepsilon}} \left(\frac{1}{\varepsilon \log 2} \right) \le \prod_{\mathcal{P} \ni p < 2^{1/\varepsilon}} \left(\frac{1}{\varepsilon \log 2} \right) .$$

Observe that the last term depends on ε, but is independent of n. Then (i) is proved. We shall now prove (ii). Let $c \ge 1$ and let $\ell = [c]$ be the integer part of c. Let $p_1 < p_2 < p_3 < \ldots$ be the prime numbers and let

$$n = n_m := (p_1 p_2 \cdots p_{\ell+1})^m$$

for every positive integer m. Then Proposition 2.10 implies

$$d(n) = (m+1)^{\ell+1} > \left(\frac{\log n}{\log(p_1 p_2 \cdots p_{\ell+1})}\right)^{\ell+1} = K(c) \log^{\ell+1}(n)$$

$$> K(c) \log^c n,$$

where $K(c) = \log^{-\ell-1}(p_1 p_2 \cdots p_{\ell+1})$ depends only on the arbitrary constant $c \geq 1$. Hence (ii) follows. $\square$

We intend to study the arithmetic mean of $d(n)$:

$$\frac{1}{R} \sum_{n \leq R} d(n).$$

A result proved by Dirichlet in 1849 shows that, in the mean, there are about $\log(R)$ divisors.

Theorem 2.14 (Dirichlet) *As $R \to +\infty$ we have*

$$\sum_{n \leq R} d(n) = R \log R + (2\gamma - 1) R + O\left(\sqrt{R}\right),$$

where γ is the Euler–Mascheroni constant.

The number γ is defined by

$$\gamma = \lim_{n \to +\infty} \left(\sum_{k=1}^{n} \frac{1}{k} - \log n\right) = 0.577215649 \cdots$$

We need a preliminary result.

Proposition 2.15 *We have*

$$\sum_{k=1}^{n} \frac{1}{k} - \log n = \gamma + O\left(\frac{1}{n}\right). \tag{2.11}$$

Proof We have

$$\sum_{k=1}^{n} \frac{1}{k} - \log n = \frac{1}{n} + \sum_{k=1}^{n-1} \frac{1}{k} - \int_1^n \frac{1}{x} dx = \frac{1}{n} + \sum_{k=1}^{n-1} \left(\frac{1}{k} - \int_k^{k+1} \frac{1}{x} dx\right).$$

Observe that the Euler–Mascheroni constant γ is the (infinite) sum of the areas of the grey parts in the following figure:

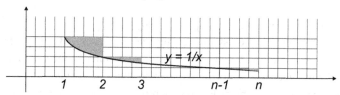

Each grey part is close to a triangle, but slightly larger. If we put all these grey parts one under the other we see that $1/2 < \gamma < 1$. Now let us prove (2.11). We have

$$0 < \gamma - \sum_{k=1}^{n-1}\left(\frac{1}{k} - \int_k^{k+1}\frac{1}{x}\,dx\right) = \sum_{k=n}^{+\infty}\left(\frac{1}{k} - \int_k^{k+1}\frac{1}{x}\,dx\right)$$

$$< \sum_{k=n}^{+\infty}\left(\frac{1}{k} - \frac{1}{k+1}\right) = \left(\frac{1}{n} - \frac{1}{n+1}\right) + \left(\frac{1}{n+1} - \frac{1}{n+2}\right) + \ldots = \frac{1}{n}.$$

$\square$

The previous result will be improved in Lemma 9.2.

Proof of Theorem 2.14   We may assume that R is an integer. Let $u = \left\lceil \sqrt{R} \right\rceil$ be the integer part of $\sqrt{R}$. Then

$$u^2 = R + O\left(\sqrt{R}\right) \tag{2.12}$$

and

$$\log u = \frac{1}{2}\log\left(R + O\left(\sqrt{R}\right)\right) = \frac{1}{2}\log R + O\left(R^{-1/2}\right). \tag{2.13}$$

In order to count the divisors of all the positive integers smaller than or equal to R, we shall consider the following figure:

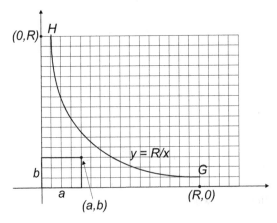

where the arc GH is part of the graph of the hyperbola $xy = R$. Assume that $a \mid k \in [1, R]$, then there exists b such that $ab = k (\leq R)$. We may see the integers a and b as the side lengths of a rectangle anchored at the origin. Since the area of this rectangle is not larger than R, then the rectangle is located 'under the graph of the hyperbola'. We identify each one of these rectangles with its upper right vertex (a, b). Therefore, $d(1) + d(2) + \ldots + d(R)$ counts the integer points (that is, the points with integer coordinates) between the graph of the hyperbola (included) and the axes (not included). Then

$$d(1) + d(2) + \ldots + d(R) = \text{card}\left\{(m_1, m_2) \in \mathbb{Z}^2 : m_1 > 0, \, m_2 > 0, \, m_1 m_2 \leq R\right\}$$

represents the area of the grey part below the graph of the hyperbola:

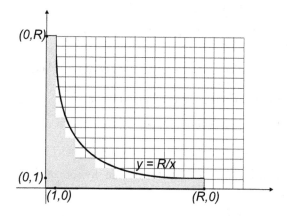

Now let $A = (0, 0)$, $B = (0, u)$, $C = (u, u)$, $D = (u, 0)$:

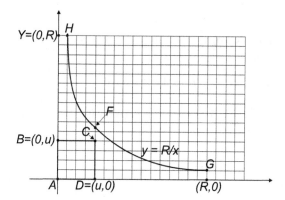

Since $(u + 1)^2 > R$, the point $(u + 1, u + 1)$ is over the graph of the hyperbola. Then the number of points between the axes and the graph of the hyperbola is twice the number of points inside the strip between AY and DF (we count the points on DF, but not the ones on AY), minus u^2 (that is, minus the number of points inside the square $ABCD$, where only half of the boundary counts). The strip between AY and DF is the union of the rectangles having base 1 and heights $\left[\frac{R}{1}\right], \left[\frac{R}{2}\right], \ldots, \left[\frac{R}{u}\right]$. Then (2.11), (2.12) and (2.13) give

$$\sum_{n \leq R} d(n) = 2 \left(\left[\frac{R}{1}\right] + \left[\frac{R}{2}\right] + \ldots + \left[\frac{R}{u}\right] \right) - u^2 = 2 \left(R + \frac{R}{2} + \ldots + \frac{R}{u} \right) - u^2 + O(u)$$

$$= 2R \left(\log u + \gamma + O\left(\frac{1}{u}\right) \right) - u^2 + O(u) = R \log R + (2\gamma - 1)R + O(\sqrt{R}) .$$

$\square$

The search for better estimates of the error

$$\sum_{n \leq R} d(n) - R \log R - (2\gamma - 1) R$$

is called the *Dirichlet divisor problem*. We shall return to it later in Chapter 9.

We now show that $\sigma(n)$ is not much larger than n.

Proposition 2.16   *For every integer $n \geq 2$ we have*

$$n + 1 \leq \sigma(n) < n + n \log n .$$

Proof   The first inequality is trivial (1 and n are divisors of n). We prove the second inequality:

$$\sigma(n) = \sum_{m \mid n} m = \sum_{m \mid n} \frac{n}{m} \leq n \sum_{m \leq n} \frac{1}{m} < n + n \int_1^n \frac{1}{x} \, dx = n + n \log n .$$

$\square$

The next result, proved by Dirichlet in 1849, shows that the average growth of $\sigma(n)$ is of order n.

Theorem 2.17 (Dirichlet)   *As $R \to +\infty$ we have*

$$\sum_{n \leq R} \sigma(n) = \frac{\pi^2}{12} R^2 + O(R \log R) .$$

Proof   By (4.35) we have

$$\sum_{n \leq R} \sigma(n) = \sum_{n \leq R} \sum_{m \mid n} m = \sum_{n \leq R} \sum_{m \mid n} \frac{n}{m} = \sum_{m \leq R} \sum_{\substack{n \leq R \\ m \mid n}} \frac{n}{m} = \sum_{m \leq R} \sum_{j \leq R/m} j$$

$$= \sum_{m \leq R} \frac{[R/m] \, ([R/m] + 1)}{2} = \frac{1}{2} \sum_{m \leq R} \left(\frac{R^2}{m^2} + O\left(\frac{R}{m}\right) \right) = \frac{R^2}{2} \sum_{m \leq R} \frac{1}{m^2} + O\left(R \sum_{m \leq R} \frac{1}{m} \right)$$

$$= \frac{R^2}{2} \sum_{m=1}^{+\infty} \frac{1}{m^2} - R^2 \sum_{m=R+1}^{+\infty} \frac{1}{m^2} + O(R \log R) = \frac{\pi^2}{12} R^2 + O(R \log R) .$$

$\square$

Gronwall proved in 1913 a very neat result on the growth of $\sigma(n)$ (see [87] or [90, Ch. XVIII]).

Theorem 2.18 (Gronwall) *Let γ be the Euler–Mascheroni constant, then*

$$\limsup_{n \to +\infty} \frac{\sigma(n)}{n \log(\log n)} = e^{\gamma} .$$

2.4 $r(n)$

Some positive integers can be written as sums of two squares: $4 = 4 + 0$, $13 = 9 + 4$, $18 = 9 + 9$, ... Some but not all: to obtain 6 we need three squares, while four squares are needed in order to obtain 7. We have a useful geometric interpretation of these two different situations: the circles centred at the origin and having radii $\sqrt{4}$, $\sqrt{13}$ or $\sqrt{18}$ meet integer points, while the circles with radii $\sqrt{6}$ or $\sqrt{7}$ do not pass by any integer point.

For every integer $n \geq 0$ the arithmetic function

$$r(n) := \operatorname{card}\left\{(a, b) \in \mathbb{Z}^2 : a^2 + b^2 = n\right\} \tag{2.14}$$

counts the number of ways to write n as a sum of two squares:

$$
\begin{aligned}
&0 = 0^2 + 0^2 && r(0) = 1 \\
&1 = (\pm 1)^2 + 0^2 = 0^2 + (\pm 1)^2 && r(1) = 4 \\
&2 = (\pm 1)^2 + (\pm 1)^2 && r(2) = 4 \\
&3 \text{ is not a sum of two squares} && r(3) = 0 \\
&4 = (\pm 2)^2 + 0^2 = 0^2 + (\pm 2)^2 && r(4) = 4 \\
&5 = (\pm 2)^2 + (\pm 1)^2 = (\pm 1)^2 + (\pm 2)^2 && r(5) = 8 \\
&6 \text{ is not a sum of two squares} && r(6) = 0 \\
&7 \text{ is not a sum of two squares} && r(7) = 0 \\
&8 = (\pm 2)^2 + (\pm 2)^2 && r(8) = 4 \\
&9 = (\pm 3)^2 + 0^2 = 0^2 + (\pm 3)^2 && r(9) = 4 \\
&10 = (\pm 3)^2 + (\pm 1)^2 = (\pm 1)^2 + (\pm 3)^2 && r(10) = 8 \\
&11 \text{ is not a sum of two squares} && r(11) = 0
\end{aligned}
$$

$$
\begin{aligned}
&\vdots && \vdots \\
&65 = (\pm 7)^2 + (\pm 4)^2 = (\pm 4)^2 + (\pm 7)^2 && \\
&\quad = (\pm 8)^2 + (\pm 1)^2 = (\pm 1)^2 + (\pm 8)^2 && r(65) = 16 \\
&\vdots && \vdots
\end{aligned}
$$

The arithmetic function $r(n)$ has an irregular behaviour, and we shall study its arithmetic mean $\frac{1}{n} \sum_{j=0}^{n} r(j)$. The following result was pointed out by Gauss in 1801.

Theorem 2.19 (Gauss)   *As $n \to +\infty$ we have*

$$\frac{1}{n} \sum_{j=0}^{n} r(j) = \pi + O\left(n^{-1/2}\right) . \tag{2.15}$$

The search for better estimates of the error $\frac{1}{n} \sum_{j=1}^{n} r(j) - \pi$ is called the *Gauss circle problem*. We shall study it more thoroughly in Section 8.3.

Proof   The arithmetic function $r(j)$ counts the integer points in $\mathbb{R}^2$ which belong to the circle[2] $S\left(0, j^{1/2}\right)$. That is, $r(j) = \text{card}\left(\mathbb{Z}^2 \cap S\left(0, j^{1/2}\right)\right)$. Then[3]

$$\frac{1}{n} \sum_{j=0}^{n} r(j) = \frac{1}{n} \text{card}\left(\mathbb{Z}^2 \cap B\left(0, n^{1/2}\right)\right) .$$

We may consider each point in the set $\mathbb{Z}^2 \cap B\left(0, n^{1/2}\right)$ as the centre of a square having sides of length 1 parallel to the axes:

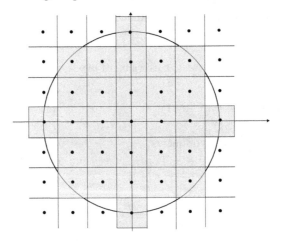

Let $K(0, n^{1/2})$ be the union of these squares (the grey set in the figure). Then

$$\text{card}\left(\mathbb{Z}^2 \cap B\left(0, n^{1/2}\right)\right) = \left|K(0, n^{1/2})\right|$$

---

[2] The sphere (circle when $d = 2$) of centre $t \in \mathbb{R}^d$ and radius R is

$$S\left(t, R\right) := \left\{u \in \mathbb{R}^d : |u - t| = R\right\} .$$

[3] The ball (disc when $d = 2$) of centre $t \in \mathbb{R}^d$ and radius R is

$$B\left(t, R\right) := \left\{u \in \mathbb{R}^d : |u - t| \leq R\right\} .$$

(where $|A|$ is the area, or volume, of a measurable set A) and

$$B\left(0, n^{1/2} - \frac{\sqrt{2}}{2}\right) \subseteq K(0, n^{1/2}) \subseteq B\left(0, n^{1/2} + \frac{\sqrt{2}}{2}\right).$$

This implies (2.15). □

2.5 $\phi(n)$

The arithmetic function $\phi(n)$ counts the integers between 1 and n which are coprime with n :

$$\phi(n) := \sum_{\substack{1 \le j \le n \\ (j,n)=1}} 1. \tag{2.16}$$

$\phi(n)$ is called the *Euler function*.

We have $\phi(1) = 1$, $\phi(2) = 1$, $\phi(3) = 2$, $\phi(4) = 2$, $\phi(5) = 4$, $\phi(6) = 2$, $\phi(7) = 6$, $\phi(8) = 4$, $\phi(9) = 6$, $\phi(10) = 4$, $\phi(11) = 10$, $\phi(12) = 4$ and $\phi(p) = p-1$ if and only if p is a prime number.

Theorem 2.20 *For every positive integer n we have*

$$\sum_{m|n} \phi(m) = n. \tag{2.17}$$

Proof After having simplified the n fractions $\frac{1}{n}, \frac{2}{n}, \ldots, \frac{n}{n}$ we obtain

$$\left\{\frac{1}{n}, \frac{2}{n}, \ldots, \frac{n}{n}\right\} = \left\{\frac{s}{m} : m \mid n, \ 0 < s \le m, \ (s,m) = 1\right\}.$$

Observe that there are $\phi(m)$ fractions with denominator m. Then (2.17) follows. □

Corollary 2.21 *The function $\phi(n)$ is multiplicative.*

Proof The function $N(n)$ is multiplicative. Then, by (2.17) and Corollary 2.6, $\phi(n)$ is a multiplicative function. □

Remark 2.22 Theorem 2.20 can be inverted as follows. If an arithmetic function f satisfies

$$\sum_{m|n} f(m) = n \tag{2.18}$$

for every n, then $f = \phi$. Indeed, (2.18) and the Möbius inversion formula (Theorem 2.8) give

$$f = \mu * N.$$

Then there is only one arithmetic function f satisfying (2.18). Hence, $f = \phi$. In other words, (2.17) provides another definition of ϕ:

$$\phi = \mu * N . \tag{2.19}$$

Corollary 2.23   *For every n we have*

$$\phi(n) = n \sum_{m|n} \frac{\mu(m)}{m} .$$

Proposition 2.24   *Let $n \geq 2$ be an integer with canonical decomposition $n = p_1^{m_1} p_2^{m_2} \cdots p_N^{m_N}$. Then*

$$\phi(n) = n \prod_{j=1}^{N} \left(1 - \frac{1}{p_j}\right) . \tag{2.20}$$

Proof   Since ϕ is multiplicative, it is enough to prove (2.20) when $n = p^s$ and p is a prime number. That is, we have to prove that

$$\phi(p^s) = p^s \left(1 - \frac{1}{p}\right) .$$

Indeed, the positive integers that are not larger than p^s and are not coprime with p are the p^{s-1} numbers $p, 2p, 3p, \ldots, p^{s-1}p$. Then

$$\phi(p^s) = p^s - p^{s-1} = p^s \left(1 - \frac{1}{p}\right) .$$

$\square$

The Dirichlet inverse of the Euler function has a quite simple form.

Theorem 2.25   *We have*

$$\phi^{*-1}(n) = \begin{cases} 1 & \text{if } n = 1, \\ \prod_{\mathcal{P} \ni p | n} (1 - p) & \text{if } n > 1. \end{cases} \tag{2.21}$$

Proof   By (2.19) we have $\phi = \mu * N$. Then $\phi^{*-1} = \mu^{*-1} * N^{*-1}$. Since

$$N * (\mu N) = \sum_{d|n} \frac{n}{d} \mu(d) d = n \sum_{d|n} \mu(d) = I(n) ,$$

we have $N^{*-1} = \mu N$. Then Lemma 2.7 implies

$$\phi^{*-1}(n) = (u * (\mu N))(n) = \sum_{d|n} d\mu(d) .$$

By Theorem 2.5 we know that ϕ^{*-1} is a multiplicative function. Then it suffices to check (2.21) when $n = p^s$ (the power of a prime number). We have

$$\phi^{*-1}(p^s) = \sum_{d|p^s} d\mu(d) = \sum_{j=0}^{s} p^j \mu(p^j) = 1 - p.$$

This implies (2.21). $\qquad\qquad\square$

We shall now consider the behaviour of $\phi(n)$ for n large.

If we reread the definitions of $\sigma(n)$ and $\phi(n)$, we see that these arithmetic functions show a sort of complementarity. Indeed, $\sigma(n)$ is large when there are many divisors, while $\phi(n)$ is large when n has few divisors. We have seen in Theorem 2.17 that $\sigma(n)$ grows, in the mean, with order n, and we shall see in Theorem 2.29 that the same happens for $\phi(n)$. The following result is coherent with the previous remarks.

Theorem 2.26 *For every positive integer n we have*

$$\frac{6}{\pi^2} < \frac{\sigma(n)\phi(n)}{n^2} \le 1. \qquad\qquad (2.22)$$

Proof For $n = 1$ the inequalities in (2.22) are true. When $n > 1$ we write its canonical decomposition $n = p_1^{m_1} p_2^{m_2} \cdots p_N^{m_N}$. Then, by Propositions 2.10 and 2.24, we have

$$\sigma(n) = \prod_{j=1}^{N} \frac{p_j^{m_j+1} - 1}{p_j - 1} = n \prod_{j=1}^{N} \frac{1 - p_j^{-m_j-1}}{1 - p_j^{-1}}$$

and

$$\phi(n) = n \prod_{j=1}^{N} \left(1 - p_j^{-1}\right).$$

So

$$\frac{\sigma(n)\phi(n)}{n^2} = \prod_{j=1}^{N} \left(1 - p_j^{-m_j-1}\right) \le 1. \qquad\qquad (2.23)$$

As for the first inequality in (2.22), we use (2.23), (1.6) and (4.35) to obtain

$$\frac{\sigma(n)\phi(n)}{n^2} = \prod_{j=1}^{N} \left(1 - p_j^{-m_j-1}\right) \ge \prod_{j=1}^{N} \left(1 - p_j^{-2}\right) > \prod_{j=1}^{+\infty} \left(1 - p_j^{-2}\right)$$

$$= \left(\sum_{k=1}^{+\infty} \frac{1}{k^2}\right)^{-1} = \frac{6}{\pi^2}.$$

$\qquad\qquad\square$

Corollary 2.27 *For every integer $n \geq 2$ we have*

$$\frac{6}{\pi^2} \frac{n}{1 + \log n} \leq \phi(n) < n .$$

Proof The first inequality follows from Theorem 2.26 and Proposition 2.16. The second inequality is trivial. $\square$

In 1903 Landau proved the following result, similar to Theorem 2.18.

Theorem 2.28 (Landau) *We have*

$$\liminf_{n \to +\infty} \frac{\phi(n) \log (\log n)}{n} = e^{-\gamma} ,$$

where γ is the Euler–Mascheroni constant.

In 1874 Mertens proved the following estimate for the arithmetic mean of Euler's ϕ function.

Theorem 2.29 (Mertens) *We have*

$$\sum_{n \leq R} \phi(n) = \frac{3}{\pi^2} R^2 + O(R \log R)$$

as $R \to +\infty$.

Proof The proof is similar to that of Theorem 2.17. By Corollary 2.23 and (2.7) we have

$$\sum_{n \leq R} \phi(n) = \sum_{n \leq R} \sum_{mm' = n} m' \mu(m) = \sum_{m \leq R} \mu(m) \sum_{m' \leq R/m} m'$$

$$= \sum_{m \leq R} \mu(m) \frac{[R/m] (1 + [R/m])}{2} = \frac{1}{2} \sum_{m \leq R} \mu(m) \left(\frac{R^2}{m^2} + O\left(\frac{R}{m}\right) \right)$$

$$= \frac{1}{2} R^2 \sum_{m \leq R} \frac{\mu(m)}{m^2} + O\left(R \sum_{m \leq R} \frac{1}{m} \right)$$

$$= \frac{1}{2} R^2 \sum_{m=1}^{+\infty} \frac{\mu(m)}{m^2} + O\left(R^2 \sum_{m > R} \frac{1}{m^2} \right) + O(R \log R)$$

$$= \frac{1}{2} R^2 \sum_{m=1}^{+\infty} \frac{\mu(m)}{m^2} + O(R \log R) = \frac{3}{\pi^2} R^2 + O(R \log R) .$$

$\square$

Remark 2.30 In a certain sense, Theorem 2.29 computes the probability that two natural numbers are coprime. Indeed, $\sum_{n \leq N} \phi(n)$ counts the pairs (m, n) of coprime integers $1 \leq m \leq n \leq N$, while $\sum_{n=1}^{N} n = N(N+1)/2$ is the total number of pairs (m, n) of integers such that $1 \leq m \leq n \leq N$. Hence the percentage of fractions m/n which cannot be simplified tends to $6/\pi^2$ as $N \to +\infty$. Indeed,

$$\lim_{N \to +\infty} \frac{\sum_{n \leq N} \phi(n)}{N(N+1)/2} = \lim_{N \to +\infty} \frac{3N^2/\pi^2 + O(N \log N)}{N(N+1)/2} = \frac{6}{\pi^2} \approx 61\% .$$

The problem discussed in the remark above can be read in a geometric way as follows (see [8, p. 63]). We may say that an integer point $(m, n) \in \mathbb{Z}^2$ is visible from the origin if the segment joining $(0, 0)$ and (m, n) contains no other integer points. Then we may say that an integer point chosen at random has probability $6/\pi^2$ of being visible from the origin (observe that if $1 \leq m \leq n$, then the point (m, n) is visible from the origin if and only if m and n are coprime).

Exercises

1) Consider the 10^n integers between 0 and $10^n - 1$, and write them in decimal form, so that they are n-digit sequences. Call *primitive* a sequence which cannot be obtained by writing two or more equal sequences one after another. Assume that, for example, $n = 12$, then 562786792002 is a primitive sequence, while 373739373739 and 373737373737 are not. We say that 373739373739 has period 6 (it comes from two sequences of 6 digits, but nothing less), while 373737373737 has period 2. If d is the period of an n-digit sequence, then $d \mid n$. Let $f(d)$ be the number of sequences with period d. Observe that $f(d)$ is independent of n. Compute $f(12)$.

2) Let f be a multiplicative arithmetic function. Prove that for every $n \in \mathbb{N}$ we have

$$\sum_{d \mid n} \mu(d) f(d) = \prod_{\mathcal{P} \ni p \mid n} (1 - f(p)) .$$

3) Let $n > 1$ and let k be the number of different prime factors of n. Prove that

$$\sum_{d \mid n} \mu^2(d) = 2^k .$$

4) Let $f(x)$ be a function defined for every real $x \geq 1$. Let

$$g(x) = \sum_{h=1}^{[x]} f\left(\frac{x}{h}\right) .$$

Prove that

$$f(x) = \sum_{k=1}^{[x]} \mu(k) g\left(\frac{x}{k}\right) .$$

5) Prove that for every positive integer n we have

$$\mu(n) = \sum_{\substack{k=0 \\ (k,n)=1}}^{n-1} e^{2\pi i k/n} .$$

6) Prove that $d(n)$ is odd if and only if n is a square.
7) Let n be a positive integer with canonical decomposition $n = 2^a p_2^{m_2} \cdots p_N^{m_N}$.
   Prove that $\sigma(n)$ is odd if and only if the numbers m_j are all even.
8) Assume that $\phi(n) \mid (n-1)$. Prove that n is square-free.
9) Assume that $\phi(n) \mid n$. Prove that n has no prime factors other than 2 and 3.
10) Prove that

$$\liminf_{n \to +\infty} \frac{\phi(n)}{n} = 0.$$

11) Prove that for every $n \in \mathbb{N}$ we have

$$\sum_{d \mid n} \frac{\mu^2(d)}{\phi(d)} = \frac{n}{\phi(n)} .$$

3

Congruences

Let $a, b \in \mathbb{Z}$ and let m be a positive integer. We say that a and b are congruent modulo m, and we write

$$a \equiv b \pmod{m},$$

if $m \mid (a - b)$. Observe that congruence modulo m is an equivalence relation. A *residue class* modulo m is a set consisting of all integers that are congruent to each other modulo m, i.e. that leave the same residue when divided by m. We usually denote a residue class by its smallest non-negative element. As an example, the numbers 3, 10, 38, -11, ... belong to the residue class 3 modulo 7. For $m \geq 2$ we denote by $\mathbb{Z}/m\mathbb{Z}$ the ring of the residue classes modulo m.

We start by describing a simple application of residue classes: the ISBN (International Standard Book Number) code, which can be found on the back cover of every fairly recent book. The code consists of 10 digits if the book was printed before January 1st, 2007, otherwise it consists of 13 digits. Let us describe the 10-digit code. Clearly the ISBN code has been created to identify each book in a unique way. As an example, the book *Trigonometric Series* by Antoni Zygmund (Cambridge University Press) has ISBN 0-521-35885-X. The code is divided into four groups. The first group identifies a country, a geographical area or a language area. In this case 0 denotes the English-speaking area. The second and third group, respectively, identify the publisher and the title. The last one is the check digit. The unexpected digit X is due to the fact that we are working modulo 11, with the eleven digits 0, 1, 2, 3, 4, 5, 6, 7, 8, 9, X. We shall see that the check digit avoids the consequences of a few common typos. Let us write an ISBN code as $a_1 a_2 \ldots a_{10}$, and let us compute

$$A = \sum_{n=1}^{10} n a_n.$$

The check digit a_{10} is chosen in order to satisfy $A \equiv 0 \pmod{11}$. For the book

Trigonometric Series we have

$$A = 1 \cdot 0 + 2 \cdot 5 + 3 \cdot 2 + 4 \cdot 1 + 5 \cdot 3 + 6 \cdot 5 + 7 \cdot 8 + 8 \cdot 8 + 9 \cdot 5 + 10 \cdot 10$$

(3.1)

$$= 330 \equiv 0 \pmod{11} .$$

A common typo consists of replacing a digit, say 3, with a different digit, say 7, and then writing the wrong code 0-521-75885-X. Hence the sum in (3.1) increases by $5 \cdot 4$, so that it is no longer $\equiv 0 \pmod{11}$. Since 11 is a prime number, there is no way to change a single digit while still preserving the congruence. Another common typo, at least for 'gronw' up 'poeple', consists of exchanging two consecutive digits. More generally, assume that we exchange two arbitrary digits, say a_2 and a_7. Then the sum A is replaced by

$$a_1 + 2a_7 + 3a_3 + 4a_4 + 5a_5 + 6a_6 + 7a_2 + 8a_8 + 9a_9 + 10a_{10}$$
$$= a_1 + 2a_2 + 3a_3 + 4a_4 + 5a_5 + 6a_6 + 7a_7 + 8a_8 + 9a_9 + 10a_{10} + 5 (a_2 - a_7)$$
$$\equiv 5 (a_7 - a_2) \not\equiv 0 \pmod{11} .$$

As a final remark, assume that we replace the condition $\sum_{n=1}^{10} n a_n \equiv 0 \pmod{11}$ with the simpler condition $\sum_{n=1}^{10} a_n \equiv 0 \pmod{11}$. As a consequence, the second typo should not be detected.

3.1 Basic properties

Addition and multiplication preserve congruences in the following sense.

Remark 3.1 Let $a_1 \equiv a_2 \pmod{m}$ and $b_1 \equiv b_2 \pmod{m}$, then $a_1 + b_1 \equiv a_2 + b_2 \pmod{m}$ and $a_1 b_1 \equiv a_2 b_2 \pmod{m}$.

The proof of the above remark is simple.

Observe that $ac \equiv bc \pmod{m}$ does not necessarily imply $a \equiv b \pmod{m}$ (as an example, we have $6 \equiv 4 \pmod{2}$, but $3 \not\equiv 2 \pmod{2}$). We have the following result.

Remark 3.2 Let $ac \equiv bc \pmod{m}$ and $d = (c, m)$, then $a \equiv b \left(\bmod \frac{m}{d} \right)$.

Proof Let us write $m = \mu d$ and $c = \gamma d$ (hence $(\mu, \gamma) = 1$). Let ℓ satisfy $ac = bc + \ell m$. Then

$$a = b + \frac{\ell m}{c} = b + \frac{\ell}{\gamma} \mu .$$

Since $(\mu, \gamma) = 1$ and $\frac{\ell}{\gamma}\mu$ is an integer, we deduce that $\gamma \mid \ell$. Then

$$a = b + \frac{\ell}{\gamma}\frac{m}{d} ,$$

that is, $a \equiv b \pmod{\frac{m}{d}}$. $\qquad\square$

As an exercise, we compute the last two digits of the number 123^{4567}, using only paper and pencil. This means that we have to find the residue class modulo 100 which contains the number 123^{4567}. The first step is to write the exponent 4567 as a sum of powers of 2 :

$$4567 = 4096 + 256 + 128 + 64 + 16 + 4 + 2 + 1 .$$

Using Remark 3.1 we can easily write (mod 100) the following powers of 123 :

$$123 \equiv 23 \pmod{100}$$
$$123^2 \equiv 23^2 \equiv 29 \pmod{100}$$
$$123^4 \equiv 29^2 \equiv 41 \pmod{100}$$
$$123^8 \equiv 41^2 \equiv 81 \pmod{100}$$
$$123^{16} \equiv 81^2 \equiv 61 \pmod{100}$$
$$123^{32} \equiv 61^2 \equiv 21 \pmod{100}$$
$$123^{64} \equiv 21^2 \equiv 41 \pmod{100} .$$

Observe that 41 has already appeared. Therefore, we know that the next terms are:

$$123^{128} \equiv 81 \pmod{100}$$
$$123^{256} \equiv 61 \pmod{100}$$
$$123^{512} \equiv 21 \pmod{100}$$
$$123^{1024} \equiv 41 \pmod{100}$$
$$123^{2048} \equiv 81 \pmod{100}$$
$$123^{4096} \equiv 61 \pmod{100} .$$

Remark 3.1 implies

$$123^{4567} = 123^{4096+256+128+64+16+4+2+1} \equiv 61 \cdot 61 \cdot 81 \cdot 41 \cdot 61 \cdot 41 \cdot 29 \cdot 23 \pmod{100} .$$

Since $61^2 \equiv 21 \pmod{100}$, $41^2 \equiv 81 \pmod{100}$, $81^2 \equiv 61 \pmod{100}$, etc. we obtain

$$123^{4567} \equiv 47 \pmod{100} .$$

We call a *complete set of residues modulo m* every set M of m integers such that if a and b are different elements in M, then $a \not\equiv b \pmod{m}$. As an example, $\{3, 11, 7, 26, 99, 2, -50\}$ is a complete set of residues modulo 7. We call a *reduced set of residues modulo m* any subset M^* of M such that every element in M^* is coprime with m. Then M^* contains $\phi(m)$ elements (see (2.16)). As an example, $\{3, 9, 87, 29\}$ is a reduced set of residues modulo 8.

Theorem 3.3 *Let $m \in \mathbb{N}$, $k \in \mathbb{Z}\setminus\{0\}$ and assume that $(k, m) = 1$.*

(i) *Let $S = \left\{x_j\right\}_{j=1}^{m}$ be a complete set of residues modulo m. Then $\left\{kx_j\right\}_{j=1}^{m}$ is also a complete set of residues modulo m.*

(ii) *Let $T = \left\{x_j\right\}_{j=1}^{\phi(m)}$ be a reduced set of residues modulo m. Then $\left\{kx_j\right\}_{j=1}^{\phi(m)}$ is also a reduced set of residues modulo m.*

Proof (i) Let $x, y \in S$ and assume that $kx \equiv ky \pmod{m}$. Then Remark 3.2 implies $x \equiv y \pmod{m}$, hence $x = y$.

(ii) Let x, y be different elements in T. We have just seen that

$$kx \not\equiv ky \pmod{m} .$$

We have $\left(x_j, m\right) = 1$ for every $j = 1, \ldots, \phi(m)$. Since $(k, m) = 1$ we deduce that $\left(kx_j, m\right) = 1$ for every j. As a consequence, $\left\{kx_j\right\}_{j=1}^{\phi(m)}$ is a reduced set of residues modulo m. $\square$

Theorem 3.4 *Let $m, n \in \mathbb{N}$ and $(m, n) = 1$.*

(i) *Let $R = \left\{x_j\right\}_{j=1}^{m}$ and $S = \{y_k\}_{k=1}^{n}$ be complete sets of residues modulo m and n, respectively. Then $\left\{nx_j + my_k\right\}_{1 \leq j \leq m, \; 1 \leq k \leq n}$ is a complete set of residues modulo mn.*

(ii) *Let $R' = \left\{x_j\right\}_{j=1}^{\phi(m)}$ and $S' = \{y_k\}_{k=1}^{\phi(n)}$ be reduced sets of residues modulo m and n, respectively. Then $\left\{nx_j + my_k\right\}_{1 \leq j \leq \phi(m), \; 1 \leq k \leq \phi(n)}$ is a reduced set of residues modulo mn.*

Proof (i) If $nx_1 + my_1 \equiv nx_2 + my_2 \pmod{mn}$, then there exists an integer h such that

$$n(x_1 - x_2) + m(y_1 - y_2) = hmn .$$

Since $(m, n) = 1$, we have $x_1 \equiv x_2 \pmod{m}$ and $y_1 \equiv y_2 \pmod{n}$.

(ii) Since ϕ is a multiplicative function and $(m, n) = 1$, we have $\phi(mn) = \phi(m)\phi(n)$. If $x \in R'$ and $y \in S'$, then

$$1 = (x, m) = (nx, m) = (nx + my, m) .$$

Indeed, we have $anx + bm = 1$ for suitable integers a and b. Hence

$$1 = a(nx + my) + (b - ay)m .$$

Then $\{nx_j + my_k\}_{1 \le j \le \phi(m), \, 1 \le k \le \phi(n)}$ is a reduced set of residues modulo mn. $\square$

For $a, b \in \mathbb{Z}$, $a \ne 0$ (we may assume actually that $a > 0$), we consider the congruence $ax \equiv b \pmod{m}$. Let x be a solution, then any $y \equiv x \pmod{m}$ is a solution. Hence the solutions (if any) of the above equation are residue classes modulo m. The example $2x \equiv 7 \pmod{4}$ shows that there may be no solutions at all. The following result tells us when solutions actually exist.

Theorem 3.5 *Let $m, a \in \mathbb{N}$ and $b \in \mathbb{Z}$. Then the equation*

$$ax \equiv b \pmod{m} \tag{3.2}$$

admits a solution if and only if $(a, m) \mid b$. There are (up to congruence modulo m) precisely (a, m) solutions. The solutions form a residue class modulo $\frac{m}{(a,m)}$.

Proof Let (3.2) have a solution, then there are integers x_0 and y_0 such that $ax_0 + my_0 = b$. This implies $(a, m) \mid b$. Conversely, let us assume that $(a, m) \mid b$ and consider the following complete set of residues modulo $\frac{m}{(a,m)}$:

$$\left\{ 0, 1, 2, \ldots, \left(\frac{m}{(a, m)} - 1 \right) \right\} .$$

We have

$$\left(\frac{a}{(a, m)}, \frac{m}{(a, m)} \right) = 1$$

then, by Theorem 3.3, we know that

$$\left\{ 0, \frac{a}{(a, m)}, \frac{2a}{(a, m)}, \ldots, \left(\frac{m}{(a, m)} - 1 \right) \frac{a}{(a, m)} \right\} \tag{3.3}$$

is a complete set of residues modulo $\frac{m}{(a,m)}$ too. There exists

$$x \in \left\{ 0, 1, 2, \ldots, \left(\frac{m}{(a, m)} - 1 \right) \right\}$$

such that

$$\frac{a}{(a, m)} x \equiv \frac{b}{(a, m)} \left(\bmod \, \frac{m}{(a, m)} \right) . \tag{3.4}$$

This implies $ax \equiv b \pmod{m}$. Then (3.2) has a solution. Finally, we show that the solutions form a residue class modulo $\frac{m}{(a,m)}$ (this implies that there are (a, m) solutions which are pairwise non-congruent modulo m). Indeed, let $y \equiv x \pmod{\frac{m}{(a,m)}}$, then y satisfies (3.4) and (3.2). We shall now show that every solution is congruent to x modulo $\frac{m}{(a,m)}$. Indeed, let y be a solution of (3.2). Then $ay \equiv ax \pmod{m}$. Remark 3.2 implies $y \equiv x \left(\mod \frac{m}{(a,m)} \right)$. $\square$

By choosing $b = 1$ in (3.2) we obtain the following consequence.

Corollary 3.6 *Let m and a be, respectively, a positive integer and a non-zero integer. Then a has an inverse $(\mod m)$ if and only if $(a, m) = 1$.*

Remark 3.7 If p is a prime number, then $\mathbb{Z}/p\mathbb{Z}$ is a field. For every integer $m \geq 2$ we write $(\mathbb{Z}/m\mathbb{Z})^{\times}$ for the multiplicative group of invertible elements in $\mathbb{Z}/m\mathbb{Z}$. This group has order $\phi(m)$ and its elements form a reduced set of residues modulo m.

3.2 The theorems of Fermat and Euler

We introduce the important Euler–Fermat theorem, proved by Euler in 1760.

Theorem 3.8 (Euler–Fermat) *Let $m \in \mathbb{N}$, $a \in \mathbb{Z}\backslash\{0\}$ and assume that $(a, m) = 1$. Then*

$$a^{\phi(m)} \equiv 1 \pmod{m} .\tag{3.5}$$

First proof Let (see Remark 3.7) $a \in (\mathbb{Z}/m\mathbb{Z})^{\times}$, and let d be the order of the group generated by a. By Lagrange's theorem on finite groups[1] we have $d \mid \phi(m)$. Then

$$a^{\phi(m)} \equiv \left(a^{d}\right)^{\phi(m)/d} \equiv 1 \pmod{m} .$$

$\square$

Second proof Let $\{r_1, r_2, \ldots, r_{\phi(m)}\}$ be a reduced set of residues modulo m. By Theorem 3.3 we know that $\{ar_1, ar_2, \ldots, ar_{\phi(m)}\}$ is also a reduced set of residues modulo m. Then there exists a permutation γ of $\{1, 2, \ldots, \phi(m)\}$ such that $r_{\gamma(k)} \equiv ar_k \pmod{m}$ for every $1 \leq k \leq \phi(m)$. By multiplying all the terms and recalling Remark 3.1, we obtain

$$r_1 r_2 \cdots r_{\phi(m)} = r_{\gamma(1)} r_{\gamma(2)} \cdots r_{\gamma(\phi(m))} \equiv (ar_1)(ar_2) \cdots (ar_{\phi(m)})$$
$$\equiv a^{\phi(m)} r_1 r_2 \cdots r_{\phi(m)} \pmod{m} .$$

Then

$$\left(a^{\phi(m)} - 1\right) r_1 r_2 \cdots r_{\phi(m)} \equiv 0 \pmod{m} .$$

Since $\left(r_1 r_2 \cdots r_{\phi(m)}, m\right) = 1$, we obtain (3.5). $\square$

---

[1] If G is a finite group, then the order (number of elements) of every subgroup of G divides the order of G.

Replacing m in (3.5) with a prime number p we obtain Fermat's little theorem, first stated by Fermat in 1640 and later proved by Euler in 1736.

Theorem 3.9 (Fermat's little theorem)   *Let $a \in \mathbb{Z}$ and let p be a prime number that does not divide a. Then*

$$a^{p-1} \equiv 1 \pmod{p} . \tag{3.6}$$

Moreover, for every integer a and every prime number p we have

$$a^p \equiv a \pmod{p} . \tag{3.7}$$

First proof   Since p is a prime number, we have $\phi(p) = p - 1$, then (3.6) follows from Theorem 3.8. Multiplying both sides in (3.6) by a, we obtain (3.7). Note that if $p \mid a$, then (3.7) is trivial. Observe that if $a \neq 0$ and $p \nmid a$, then (3.6) follows from (3.7) and Remark 3.2.   □

Second proof   It is enough to prove (3.7). We may assume that $a > 0$ and proceed by induction on a. When $a = 1$, then (3.7) is true. Let $0 < k < p$, then

$$\binom{p}{k} \equiv 0 \pmod{p} .$$

This implies

$$(a + 1)^p = \sum_{k=0}^{p} \binom{p}{k} a^k \equiv a^p + 1 \pmod{p} ,$$

$$(a + 1)^p - (a + 1) \equiv a^p - a \pmod{p} . \tag{3.8}$$

Assuming that (3.7) is true for a, we deduce it for $a + 1$.   □

Fermat's little theorem implies the following result of Euler.

Theorem 3.10 (Euler)   *There are infinitely many prime numbers of the form $4k + 1$.*

Proof   Let $n > 1$ and $N = (n!)^2 + 1$. Let p be a prime number which divides N. Then $p > n$ (otherwise we would have $p \mid n!$, which contradicts $p \mid N$) and $(n!)^2 \equiv -1 \pmod{p}$. Since $p \neq 2$, Remark 3.1 and Fermat's little theorem imply

$$1 \equiv (n!)^{p-1} = \left((n!)^2\right)^{(p-1)/2} \equiv (-1)^{(p-1)/2} \pmod{p} .$$

Then $(p - 1)/2$ is an even number, that is, $p \equiv 1 \pmod{4}$. Since n is an arbitrary positive integer and $p > n$, we have proved the existence of infinitely many prime numbers of the form $4k + 1$.   □

Remark 3.11 The existence of infinitely many prime numbers of the form $4k - 1$ is a more elementary result than that of the case $4k + 1$. Its proof is close to Euclid's proof of the infinitude of prime numbers. Indeed, let us assume that the set $F = \{3, 7, 11, 19, \ldots, p_N\}$ consists of all prime numbers of the form $4k - 1$. Let

$$P = (4 \cdot 3 \cdot 7 \cdot 11 \cdot 19 \cdots p_N) - 1 .$$

Then P is of the form $4k - 1$ and cannot be divided by any element in F. Hence all the prime divisors of P are of the form $4k + 1$. This is impossible, since

$$(4h + 1)(4\ell + 1) = 4(4h\ell + h + \ell) + 1$$

is of the form $4k + 1$.

A famous result proved by Dirichlet in 1837 (see e.g. [161]) states that if a and b are coprime numbers, then there are infinitely many prime numbers of the form $ak + b$. Dirichlet's theorem is a deep result; easy arguments like the previous ones only work for suitable choices of a and b (see [125]).

Let us now show a possible curious application (see [6]) of Fermat's and Euler's theorems. Suppose we have a deck of cards which we are going to shuffle. Shuffling is a procedure we use to randomize the deck in order to provide an element of chance in card games. Everybody knows that we need to shuffle the cards several times. We are going to show that a deck may not necessarily be well mixed even after many shuffles.

Let us start with a deck of 52 cards and define the *(perfect) shuffle*. In a perfect shuffle we have to split the deck into two parts of equal size (a bottom half and a top half) and then interweave each half as follows:

1	2	3	4	5	6	...	47	48	49	50	51	52
						$\downarrow$						
27	1	28	2	29	3	...	50	24	51	25	52	26

We claim that after shuffling the deck 52 times we will be back to the original order. Let us shuffle the deck once: the card at position y_0 ($1 \leq y_0 \leq 52$) shifts to position y_1, where $1 \leq y_1 \leq 52$ and $y_1 \equiv 2y_0 \pmod{53}$. Indeed, this is obvious if $1 \leq y_0 \leq 26$ (because in this case $y_1 = 2y_0$), while for $27 \leq y_0 \leq 52$ we observe that y_0 loses 26 positions (because we have taken away the first half of the deck), then it doubles the new position (because we have shuffled) and finally it loses 1 position (because the bottom half of the deck is 'over' the top half). That is to say,

$$y_1 = (y_0 - 26)\, 2 - 1 \equiv 2y_0 \pmod{53} \ .$$

If we shuffle n times, the card in the original position y_0 will move to the position y_n, where $1 \leq y_n \leq 52$ and $y_n \equiv 2y_{n-1} \pmod{53}$. Then $y_n \equiv 2^n y_0 \pmod{53}$. Indeed,

$$y_n = 2y_{n-1} + 53h = 2\,(2y_{n-2} + 53k) + 53h = 2^2 y_{n-2} + 53\ell = \ldots$$

Hence the deck goes back to its original order if n satisfies

$$2^n y_0 \equiv y_0 \pmod{53} \tag{3.9}$$

for every $1 \leq y_0 \leq 52$. Since $(y_0, 53) = 1$, Remark 3.2 tells us that (3.9) is equivalent to $2^n \equiv 1 \pmod{53}$. Since 53 is a prime number, Fermat's little theorem implies $2^{52} \equiv 1 \pmod{53}$. Therefore 52 perfect shuffles bring the deck of cards back to its original order.

From a more general point of view, let us consider a deck of $2m$ cards. We look for k such that $2^k y_0 \equiv y_0 \pmod{(2m + 1)}$ for every $1 \leq y_0 \leq 2m$. Since $(2, 2m + 1) = 1$, Theorem 3.8 implies

$$2^{\phi(2m+1)} \equiv 1 \pmod{(2m + 1)} \ .$$

Then $2^{\phi(2m+1)} y_0 \equiv y_0 \pmod{(2m + 1)}$ for every y_0. This means that after $\phi\,(2m + 1)$ perfect shuffles the deck of $2m$ cards goes back to its original order. In particular, if $2m + 1$ is a prime number, we have $\phi(2m + 1) = 2m$, while for, say, 8 cards we have $\phi\,(9) = 6$ shuffles.

Note that we have defined a 'perfect shuffle'. The way real people shuffle real cards is a different and very interesting problem. See [11].

We now go back to the equation $ax \equiv b \pmod{m}$.

Remark 3.12 If we assume that $(a, m) \mid b$, then the Euclidean algorithm gives us the solution to the equation

$$ax \equiv b \pmod{m} \ . \tag{3.10}$$

Indeed, as in Remark 1.5, we find λ and μ such that

$$\lambda a + \mu m = (a, m) \ .$$

Since $(a, m) \mid b$ we can choose s such that

$$b = s\,(a, m) = s\lambda a + s\mu m \ .$$

Then $a(s\lambda) \equiv b \pmod{m}$. Moreover, if $(a, m) = 1$ we can solve (3.10) by multiplying both sides by $a^{\phi(m)-1}$. Indeed, we obtain

$$a^{\phi(m)}x \equiv a^{\phi(m)-1}b \pmod{m} .$$

Then the Euler–Fermat theorem implies

$$x \equiv a^{\phi(m)-1}b \pmod{m} .$$

Let us now deduce Wilson's theorem, which appeared without proof in 1770 in a work written by his teacher Waring. The first proof appeared in the same year and was due to Lagrange.

Theorem 3.13 (Wilson) *We have*

$$p \in \mathcal{P} \Longleftrightarrow (p-1)! \equiv -1 \pmod{p} . \tag{3.11}$$

Proof Let $p = mn$ be a composite number $(1 < m < p)$. Hence $m \mid (p-1)!$ and $m \nmid ((p-1)! + 1)$. Then $p \nmid ((p-1)! + 1)$, that is,

$$(p-1)! \not\equiv -1 \pmod{p} .$$

Conversely, let $p \in \mathcal{P}$. If $p = 2$ or $p = 3$, then (3.11) is true. If $p \geq 5$, choose an integer $1 \leq x \leq p - 1$. Then Corollary 3.6 implies the existence of a unique x' such that $xx' \equiv 1 \pmod{p}$. The equation $x^2 \equiv 1 \pmod{p}$ has the unique solutions $x \equiv \pm 1 \pmod{p}$ (indeed the above congruence can be written as $(x-1)(x+1) \equiv 0 \pmod{p}$). Observe that the $p-3$ numbers $2, 3, \ldots, p-2$ are coprime with p, therefore, by Theorem 3.5, they can be subdivided into $\frac{p-3}{2}$ pairs $\left(a_i, a_i'\right)$ such that $a_i \neq a_i'$ and $a_i a_i' \equiv 1 \pmod{p}$. We deduce that

$$(p-1)! = (p-1) \cdot 2 \cdot 3 \cdots (p-2) \equiv p - 1 \equiv -1 \pmod{p} .$$

$\square$

3.3 Almost prime numbers

Fermat's little theorem states that an odd prime number n satisfies

$$2^{n-1} \equiv 1 \pmod{n} . \tag{3.12}$$

Observe that there are composite numbers which satisfy (3.12). Indeed[2]

$$2^{340} \equiv 1 \pmod{341} ,$$

---

[2] This congruence can easily be verified. Indeed, we have $2^{10} = 1024 \equiv 1 \pmod{341}$. Then

$$2^{340} = \left(2^{10}\right)^{34} \equiv 1 \pmod{341} .$$

but $341 = 11 \cdot 31$.

A particular case of a result proved by Cipolla in 1904 (see [55, 145]) shows that there are infinitely many composite numbers which satisfy (3.12).

Theorem 3.14 (Cipolla)   *For every prime number $p > 3$, the number*

$$c = (2^p - 1)\frac{2^p + 1}{3}$$

satisfies $2^{c-1} \equiv 1 \pmod{c}$.

Proof   Since

$$\frac{2^p + 1}{2 + 1} = 2^{p-1} - 2^{p-2} + \ldots + 1$$

we see that c is an odd composite integer. Fermat's little theorem and Remark 3.2 imply

$$3c = 4^p - 1 \equiv 3 \pmod{p} , \quad c \equiv 1 \pmod{p} .$$

Since c is odd we have $c \equiv 1 \pmod{2p}$. We write $c = 1 + 2\ell p$. Since

$$2^{2p} = 3c + 1 \equiv 1 \pmod{c}$$

we have

$$2^{c-1} = \left(2^{2p}\right)^{\ell} \equiv 1 \pmod{c} .$$

$\square$

A result proved by Erdős[3] in [74] shows that the odd composite numbers n which satisfy $2^{n-1} \equiv 1 \pmod{n}$ are 'little o' of the prime numbers.

Definition 3.15   An odd, composite number n which satisfies (3.12) is called an *almost prime number*. We write C for the set of almost prime numbers.

Theorem 3.16 (Erdős)   *For every real positive x, let $C_x = \{n \in C : n \leq x\}$. Then for x large enough we have*

$$\operatorname{card} C_x < xe^{-\frac{1}{3}\sqrt[4]{\log x}} . \tag{3.13}$$

---

[3] Paul Erdős (Budapest 1913, Warsaw 1996) has been one of the greatest mathematicians of the twentieth century. He has written hundreds of papers with hundreds of colleagues. The *Erdős number* (see www.oakland.edu/enp/) was created by his friends and it describes the 'collaborative distance' between Erdős and the other mathematicians. Erdős has Erdős number 0. If you wrote a paper with Erdős you have Erdős number 1. If you do not have Erdős number 1, but you wrote a paper with a colleague who has Erdős number 1, then you have Erdős number 2. And so on. The author of this book has Erdős number 3. For a biography of Paul Erdős see [97]. Besides the following theorem on almost prime numbers, we shall see the Erdős–Turán inequality and the Erdős–Fuchs theorem.

Remark 3.17   By Theorem 1.15 we have $\pi(x) \geq c \frac{x}{\log x}$. Then (3.13) implies

$$\lim_{x \to +\infty} \frac{\operatorname{card} C_x}{\pi(x)} \leq c_1 \lim_{x \to +\infty} \frac{\log x}{e^{\frac{1}{3} \sqrt[4]{\log x}}} = c_1 \lim_{y \to +\infty} \frac{81 y^4}{e^y} = 0 .$$

Then $\operatorname{card} C_x = o\,(\pi(x))$ as $x \to +\infty$.

Proof of Theorem 3.16   For every $n \in C$ let

$$g(n) = \min \{s \in \mathbb{N} : 2^s \equiv 1 \pmod n\}$$

(since $2^{n-1} \equiv 1 \pmod n$, we know that the set $\{s \in \mathbb{N} : 2^s \equiv 1 \pmod n\}$ is not empty). Let us write the disjoint union

$$C_x = C_x^1 \cup C_x^2 ,$$

where

$$C_x^1 = \left\{ n \in C_x : g(n) \leq e^{\sqrt{\log x}} := H \right\} .$$

Observe that $n \in C_x^1$ implies

$$n \mid P := \prod_{r=1}^{[H]} (2^r - 1) . \tag{3.14}$$

Let $q_1, q_2, \ldots, q_k$ be the prime factors of P. Then

$$C_x^1 \subseteq \Gamma_1 := \{n \in \mathbb{N} : n \leq x \text{ and } n \text{ has no prime factors}$$
$$\text{other than } q_1, q_2, \ldots, q_k\} .$$

Observe that every factor $2^r - 1$ in (3.14) has less than r prime factors, otherwise by multiplying them we should have $2^r - 1 \geq (r+1)!$ (which is false for every $r \in \mathbb{N}$). This implies

$$k < \sum_{r=1}^{[H]} r \leq H^2 .$$

Let us write the disjoint union

$$\Gamma_1 = \Gamma_{1,1} \cup \Gamma_{1,2} ,$$

where

$$\Gamma_{1,1} := \left\{ n \in \Gamma_1 : n \text{ has less than } W := \frac{1}{10} \sqrt{\log x} \right.$$
$$= \frac{1}{10} (\log H) \text{ prime factors} \right\} .$$

In order to estimate card $\Gamma_{1,1}$ we observe that if $\Gamma_{1,1} \ni m \leq x$ and $g^\alpha \mid m$ (and if g is a prime number), then

$$\alpha \leq \log_g m \leq \frac{\log x}{\log 2}.$$

Then every power of g contributes to card $\Gamma_{1,1}$ with at most $\left(\frac{\log x}{\log 2} + 1\right)$ choices (we have to consider g^0 also). We have at most W choices for g and, therefore, if we write $x = e^{100y^2}$ we have

$$\operatorname{card} \Gamma_{1,1} \leq \left(\frac{\log x}{\log 2} + 1\right)^W = \left(\frac{\log x}{\log 2} + 1\right)^{\frac{1}{10}\sqrt{\log x}}$$

$$= \left(\frac{100y^2}{\log 2} + 1\right)^y \leq e^{25y^2} = \sqrt[4]{x}$$

for x large enough. Now let $m \in \Gamma_{1,2}$ (hence m has at least W prime factors). We write $d(m)$ and $v(m)$ for the number of divisors of m and the number of prime factors of m, respectively. Then

$$d(m) \geq 2^{v(m)} > 2^W \geq e^{W/2} \tag{3.15}$$

(note that if m is square-free, then $d(m) = 2^{v(m)}$). By Theorem 2.14 we have

$$\sum_{m \leq x} d(m) = x \log x + (2\gamma - 1)x + O\left(\sqrt{x}\right).$$

Then (3.15) implies

$$2x \log x > \sum_{m \in \Gamma_{1,2}} d(m) \geq (\operatorname{card} \Gamma_{1,2}) e^{W/2}$$

for large x. Hence

$$\operatorname{card} \Gamma_{1,2} < (2x \log x) e^{-\frac{1}{20}\sqrt{\log x}}.$$

Then, for large x we have

$$\operatorname{card} C_x^1 < x^{1/4} + (2x \log x) e^{-\frac{1}{20}\sqrt{\log x}} < x e^{-\frac{1}{30}\sqrt{\log x}}.$$

We now pass to C_x^2 and write it as the disjoint union

$$C_x^2 = C_{2,1} \cup C_{2,2},$$

where

$$C_{2,1} \tag{3.16}$$

$$= \left\{ n \in C_x^2 : \exists p \in \mathcal{P} \text{ such that } p \mid n \text{ and } \delta := (n-1, p-1) \geq e^{\sqrt[4]{\log x}} := T \right\}.$$

Since n is a composite number, we have $n = pm$ with $m > 1$. Since $m \le x/p$, (3.16) implies the existence of at most $x/(\delta p)$ choices for m. Indeed, if m_0 is one of these choices (that is, $m_0 \le x/p$ and $(pm_0 - 1, p - 1) = \delta$), then we claim that $m_0 + j$ is not a choice for m if $0 < |j| < \delta$. Indeed, for suitable ℓ and t we have

$$ pm_0 - 1 = \delta\ell , \qquad p - 1 = \delta t . \tag{3.17} $$

The assumption

$$ p(m_0 + j) - 1 = \delta r $$

would therefore imply

$$ pm_0 + (1 + \delta t)\, j - 1 = \delta r , $$
$$ pm_0 - 1 = -j + \delta\,(r - jt) , $$

against (3.17). Then we have at most $x/(\delta p)$ choices for m. Since $\delta \mid (p - 1)$ and $T = e^{\sqrt[4]{\log x}}$ we obtain

$$ \operatorname{card} C_{2,1} \le \sum_{\substack{\delta \ge T}} \sum_{\substack{p,m \\ pm \le x \\ \delta \mid (p-1)}} 1 \le \sum_{\delta \ge T} \sum_{\substack{p,t \\ p \le x \\ p = \delta t + 1}} \frac{x}{\delta p} < x \sum_{\delta \ge T} \sum_{t < x/\delta} \frac{1}{\delta^2 t} < 2x \sum_{\delta \ge T} \frac{1}{\delta^2} \log\left(\frac{x}{\delta}\right) $$

$$ < 2x \log x \sum_{\delta \ge T} \frac{1}{\delta^2} < \frac{3x \log x}{T} = \frac{3x \log x}{e^{\sqrt[4]{\log x}}} < x e^{-\frac{1}{2}\sqrt[4]{\log x}} $$

for x large enough. Finally we study

$$ C_{2,2} = \left\{ n \in C_2^x : n = p_1^{\alpha_1} p_2^{\alpha_2} \cdots p_v^{\alpha_v} \right. \tag{3.18} $$
$$ \left. \text{and } \delta_j := \left(n - 1, p_j - 1 \right) < T \text{ for every } j \right\} . $$

Since n is an almost prime number, we have $2^{n-1} \equiv 1 \pmod{n}$ and then[4] $g(n) \mid (n - 1)$. In the same way, the Fermat–Euler theorem (see (3.5)) implies $g(n) \mid \phi(n)$. Hence $g(n) \mid (n - 1, \phi(n))$ and then $g(n) \le (n - 1, \phi(n))$. Since $H < g(n)$ for every $n \in C_{2,2}$, then (3.18) and Proposition 2.24 imply

$$ e^{\sqrt{\log x}} = H < g(n) \le (n - 1, \phi(n)) = \left(n - 1, n \prod_{j=1}^{v} \left(\frac{p_j - 1}{p_j} \right) \right) $$

---

[4] Observe that $2^a \equiv 1 \pmod{n}$ implies $g(n) \mid a$. Indeed, if $a \ge g(n)$ and assuming $a = \ell g(n) + r$ with $0 < r < g(n)$, we have

$$ sn = 2^a - 1 = 2^r \left(2^{g(n)} \right)^\ell - 1 = 2^r \left(2^{g(n)} - 1 + 1 \right)^\ell - 1 $$
$$ = 2^r (wn + 1)^\ell - 1 = 2^r (1 + hn) - 1 = 2^r + un - 1 . $$

Then $2^r \equiv 1 \pmod{n}$, but this cannot happen for $r < g(n)$.

$$= \left(n - 1, \prod_{j=1}^{v} p_j^{\alpha_j - 1} \prod_{j=1}^{v} (p_j - 1)\right) = \left(n - 1, \prod_{j=1}^{v} (p_j - 1)\right)$$

$$\leq \prod_{j=1}^{v} (n - 1, p_j - 1) \leq T^v ,$$

because for every j we have $(n - 1, p_j) = 1$ and because $(a, bc) \leq (a, b)(a, c)$. Then

$$v = v(n) \geq \log_T H = \frac{\log H}{\log T} = \frac{\sqrt{\log x}}{\sqrt[4]{\log x}} = \log T . \qquad (3.19)$$

Again, let $d(n)$ count the divisors of n. Then Theorem 2.14, an argument similar to that in (3.15) and (3.19) give, for large x,

$$2x \log x > \sum_{n \leq x} d(n) \geq \sum_{n \leq x} 2^{v(n)} \geq \sum_{n \in C_{2,2}} 2^{v(n)} \geq 2^{\log T} \text{ card } C_{2,2} .$$

Hence

$$\text{card } C_{2,2} \leq 2x \log x \, e^{-(\log 2) \log T} = 2x \log x \, e^{-(\log 2) \sqrt[4]{\log x}} \leq x e^{-\frac{1}{2} \sqrt[4]{\log x}} .$$

By collecting the above estimates we end the proof. □

We can refine the above primality test as follows.

Fermat's little theorem says that if n is a prime number and n is coprime with a, then

$$a^{n-1} \equiv 1 \pmod{n} . \qquad (3.20)$$

It is natural to ask whether there exists a composite number n which satisfies (3.20) for every a coprime with n. These numbers do indeed exist and they are called Carmichael numbers after Carmichael who discovered, in 1910, the smallest number with this property: 561.

It is not difficult to show that $561 = 3 \cdot 11 \cdot 17$ is a Carmichael number. Indeed, assume $(a, 561) = 1$. Then

$$3 \nmid a , \quad 11 \nmid a , \quad 17 \nmid a$$

and Fermat's little theorem implies

$$a^2 \equiv 1 \pmod 3 , \quad a^{10} \equiv 1 \pmod{11} , \quad a^{16} \equiv 1 \pmod{17} .$$

Observe that 2, 10 and 16 divide 560. Then

$$a^{560} \equiv 1 \pmod 3 , \quad a^{560} \equiv 1 \pmod{11} , \quad a^{560} \equiv 1 \pmod{17} .$$

Whence $a^{560} \equiv 1 \pmod{561}$. Then 561 is a Carmichael number.

In 1994 Alford, Granville and Pomerance proved the existence of infinitely many Carmichael numbers (see [5]).

3.4 The Chinese remainder theorem

Let us start by describing a problem which leads us to the study of a system of linear congruences.

Bob is sick and next Monday he will start taking *medicine* α every 3 days. Next Tuesday he will start taking *medicine* β every 2 days, and next Thursday he will start taking *medicine* γ every 5 days. Since Wednesdays are heavy days for him, Bob wishes to know if and when he will have to take all three medicines on Wednesday. Let us write x for the first of these Wednesdays (if any), and let us count the days as follows. Monday of the first week is Day 1, Tuesday of the first week is Day 2, ... , Tuesday of the second week is Day 9, ... Then x must satisfy

$$x = 1 + 3k_1 = 2 + 2k_2 = 4 + 5k_3 = 3 + 7k_4$$

for suitable integers k_1, k_2, k_3, k_4. That is to say

$$\begin{cases} x \equiv 1 \pmod 3 \\ x \equiv 2 \pmod 2 \\ x \equiv 4 \pmod 5 \\ x \equiv 3 \pmod 7 . \end{cases}$$

Such an x exists because of the *Chinese remainder theorem*.[5]

Theorem 3.18 (Chinese remainder theorem) *Let* $m_1, m_2, \ldots, m_n$ *be pairwise coprime positive integers, and let* $M = m_1 m_2 \cdots m_n$. *Then, for every choice of*

---

[5] The name has its roots in the third century, when the Chinese mathematician Sun Zi found a way to compute an integer x having residues $2, 3, 2$ when divided respectively by $3, 5, 7$, that is,

$$\begin{cases} x \equiv 2 \pmod 3 \\ x \equiv 3 \pmod 5 \\ x \equiv 2 \pmod 7 . \end{cases} \tag{3.21}$$

Indeed, he pointed out that

$$\begin{cases} 5 \mid 70 , \quad 7 \mid 70 , \quad 70 \equiv 1 \pmod 3 \\ 3 \mid 21 , \quad 7 \mid 21 , \quad 21 \equiv 1 \pmod 5 \\ 3 \mid 15 , \quad 5 \mid 15 , \quad 15 \equiv 1 \pmod 7 . \end{cases}$$

Then $233 = 2 \cdot 70 + 3 \cdot 21 + 2 \cdot 15$ satisfies (3.21). Since we are interested in a solution modulo 105, we may choose $x = 23$.

$b_1, b_2, \ldots, b_n \in \mathbb{Z}$, the system

$$
\begin{cases}
x \equiv b_1 \pmod{m_1} \\
 \vdots \\
x \equiv b_n \pmod{m_n}
\end{cases}
\tag{3.22}
$$

has a unique (mod M) solution.

Proof The argument is precisely that described in the historical footnote. For every $j = 1, \ldots, n$ let $M_j = M/m_j$. Then $(M_j, m_j) = 1$ for every j and Theorem 3.5 implies the existence of an integer N_j such that $0 \le N_j < m_j$ and $M_j N_j \equiv 1 \pmod{m_j}$. We also have $M_j N_j \equiv 0 \pmod{m_k}$ for every $k \ne j$, therefore the number

$$
x = \sum_{j=1}^{n} b_j M_j N_j
$$

satisfies all the congruences in (3.22). In order to prove uniqueness, let y be a solution of all the congruences in (3.22). Then $y \equiv x \pmod{m_j}$ for every $j = 1, \ldots, n$, hence $y \equiv x \pmod{M}$. □

See [166] for a different formulation of the Chinese remainder theorem.

Let us now go back to the three medicines α, β, γ. Let

$$
x = \sum_{j=1}^{4} b_j M_j N_j \, ,
$$

where

$$
b_1 = 1 \, , \quad b_2 = 2 \, , \quad b_3 = 4 \, , \quad b_4 = 3 \, ,
$$

$$
M_1 = 70 \, , \quad M_2 = 105 \, , \quad M_3 = 42 \, , \quad M_4 = 30
$$

and

$$
70 N_1 \equiv 1 \pmod 3 \, , \quad 105 N_2 \equiv 1 \pmod 2 \, , \tag{3.23}
$$
$$
42 N_3 \equiv 1 \pmod 5 \, , \quad 30 N_4 \equiv 1 \pmod 7 \, .
$$

The numbers

$$
N_1 = 1 \, , \quad N_2 = 1 \, , \quad N_3 = 3 \, , \quad N_4 = 4
$$

satisfy (3.23). Then

$$
x = 70 + 210 + 504 + 360 = 1144 \equiv 94 \pmod{210} \, .
$$

Since $94 = 3 + 7 \cdot 13$, Bob will not have to take the three medicines on Wednesday before the 13th week.

At the end of the last chapter we saw that an integer point chosen at random has probability $6/\pi^2$ of being visible from the origin. We shall apply here the Chinese remainder theorem to show that there are arbitrarily large 'squares' which are not visible from the origin (see [8]).

Proposition 3.19 *For every positive integer N there exists $(a, b) \in \mathbb{Z}^2$ such that the set*

$$C = \{(a + j, b + k) : 0 \le j, k \le N, \ j, k \in \mathbb{Z}\}$$

(that is, the intersection of $\mathbb{Z}^2$ with a closed square of side length N) is not visible from the origin, that is, there are no points $(p, q) \in C$ such that p and q are coprime numbers.

Proof We arrange the first N^2 prime numbers $p_1, p_2, \ldots, p_{N^2}$ into the following $N \times N$ table:

$$\begin{bmatrix} p_1 & p_2 & \cdots & p_N \\ p_{N+1} & p_{N+2} & \cdots & p_{2N} \\ \vdots & \vdots & \ddots & \vdots \\ p_{N^2-N+1} & p_{N^2-N+2} & \cdots & p_{N^2} \end{bmatrix}.$$

Let

$$m_1 = p_1 p_2 \cdots p_N,$$
$$m_2 = p_{N+1} p_{N+2} \cdots p_{2N},$$
$$\vdots$$
$$m_N = p_{N^2-N+1} p_{N^2-N+2} \cdots p_{N^2}.$$

Note that $m_1, m_2, \ldots, m_N$ are pairwise coprime. Let us repeat the previous argument, letting $M_1, M_2, \ldots, M_N$ be the products of the columns of the table. By the Chinese remainder theorem, the two systems

$$\begin{cases} x \equiv -1 \pmod{m_1} \\ x \equiv -2 \pmod{m_2} \\ \vdots \\ x \equiv -N \pmod{m_N} \end{cases} \qquad \begin{cases} y \equiv -1 \pmod{M_1} \\ y \equiv -2 \pmod{M_2} \\ \vdots \\ y \equiv -N \pmod{M_N} \end{cases}$$

have solutions a and b (unique modulo $M_1 M_2 \cdots M_N = m_1 m_2 \cdots m_N$). Now let

$(a + j, b + k)$ be any point in C. Then

$$a \equiv -j \ (\mathrm{mod}\, m_j)\,, \qquad b \equiv -k \ (\mathrm{mod}\, M_k)\,.$$

Whence the prime number belonging to the jth row and the kth column divides $a + j$ and $b + k$. Then $(a + j, b + k)$ is not visible from the origin. □

3.5 The RSA encryption

Suetonius[6] described the method used by Caesar to hide his private messages as follows. Each letter would be replaced by the letter three positions down the alphabet: A was replaced by D, B by E, ... as in the following table, with the English alphabet as reference:

plaintext	A	B	C	D	E	$\cdots$	V	W	X	Y	Z
ciphertext	D	E	F	G	H	$\cdots$	Y	Z	A	B	C

For example, the phrase

TRY CAESAR CIPHER

would become

WUB FDHVDU FLSKHU

Caesar's cipher contained the typical elements of an encryption: the *algorithm* and the *key*. The algorithm is the rule used to pass from the plaintext to the ciphertext. Usually this rule depends on a parameter, the key. In Caesar's cipher the algorithm is the shift of the letters, while the key is 3. There are few non-trivial ways (25) to shift the letters, so it is not difficult to break Caesar's cipher by trying out all the keys. We can improve Caesar's cipher by permuting the letters. This can be done in about 26! non-trivial ways, so that it is impossible to try all the keys. This encryption had been introduced in India, but it became unsafe when the Arabs started using a statistical argument to break it: the frequency analysis. This system implies knowledge of the frequency of each letter in the language used for the message (see e.g. Conan Doyle's *The adventure of the dancing men* or Poe's *The gold bug*). If the ciphertext is long

---

[6] 'Extant et ad Ciceronem, item ad familiares domesticis de rebus, in quibus, si qua occultius perferenda erant, per notas scripsit, id est sic structo litterarum ordine, ut nullum verbum effici posset; quae si qui investigare et persequi velit, quartam elementorum litteram, id est D pro A et perinde reliquas commutet.' (De vita Caesarum, Divus Iulius §56)

enough, one can make reasonable guesses on the most frequent symbols, and then try to decipher the whole message.

This was the beginning of the war between the cryptographers (those who discovered new algorithms) and the cryptanalysts (those who tried to break the codes). More sophisticated encryptions had to be created, and at the same time cryptography became a very serious issue. This happened when both transmitting messages and intercepting them (e.g., by radio) became easier. In 1914, in Tannenberg, a Russian army was completely destroyed by German troops, and one of the reasons was that Russian messages were transmitted in the clear.

After World War I, the German engineer Scherbius invented the most famous cipher machine in history: Enigma. This name actually describes a family of electromechanical machines that underwent constant development for over 20 years. In the early 1930s the Polish (mainly Rejewski) were the first to crack the Enigma code. This was a turning point in the history of cryptography, since for the first time cryptanalysts were found not among linguists, but rather among mathematicians. Before the war the Polish breakthrough was the starting point for the British codebreaking centre, where the genius of Turing played a central role. At the beginning of 1940 British cryptanalysts could read the messages written using Enigma. See [157] for a fascinating history of cryptography.

During the first half of the twentieth century, cryptographers and cryptanalysts (partially) replaced paper and pencil with electromechanical machines. In the second half of the century it was the turn of computers. This change came together with a fundamental theoretical and practical step, the use of two keys: a *public key* which everybody can use to send an encrypted message, and a *private key* which is known only to the addressee, and is necessary to decrypt the message. In other words, it is easy to turn a plaintext into a cyphertext, but the converse is difficult: an encryption system does not need to be symmetric any longer. In fact, symmetric encryption systems are still used nowadays when the transmission of the key is not a problem, or when they can be used together with a public key system.

The most famous public key system was created in 1978 by Rivest, Shamir and Adleman from MIT and is called RSA.[7] In the RSA system we consider two different large prime numbers p and q, together with their product $n = pq$. Let ϕ be the Euler function, then

$$\phi(n) = \phi(p)\phi(q) = (p-1)(q-1) = n + 1 - p - q \, .$$

We then choose h such that $(h, \phi(n)) = 1$. Now we consider a message and

---

[7] The name comes from the initials of the authors. It is RSA and not ARS because Adleman regarded his contribution as minimal (see [148]).

write it in a natural way as a number. As an example, we can put $A = 01$, $B = 02, C = 03, \ldots$ We have 100 numbers between 00 and 99 and they suffice to represent letters, numbers and punctuation. The message is now a number with many digits and we split this number into blocks, so that each one of them is a number $\leq n$. Let P be one of these blocks. We switch from the plaintext P to the ciphertext C uniquely defined by

$$\begin{cases} C \equiv P^h \pmod{n}, \\ 0 \leq C < n. \end{cases} \tag{3.24}$$

Now let us show how to go back to P. First we find d such that

$$dh \equiv 1 \pmod{\phi(n)} \tag{3.25}$$

(d exists by Theorem 3.5, and we also know how to compute it, see Remark 3.12). We write $dh = 1 + \ell\phi(n)$ and claim that

$$C^d \equiv P \pmod{n} . \tag{3.26}$$

Indeed we have $\pmod{n}$

$$C^d \equiv \left(P^h\right)^d = P^{1+\ell(p-1)(q-1)} .$$

Assume that $p \nmid P$. Then Fermat's little theorem gives $P^{p-1} \equiv 1 \pmod{p}$, so that

$$C^d \equiv P\left(P^{p-1}\right)^{\ell(q-1)} \equiv P \pmod{p} .$$

If $p \mid P$, then

$$C^d \equiv P^{hd} \equiv 0 \equiv P \pmod{p} .$$

Hence $C^d \equiv P \pmod{p}$ for every P. In the same way we have $C^d \equiv P \pmod{q}$. Since p and q are different prime numbers, (3.26) is proved.

At this point it is perfectly safe to let the whole world know h and n, so that everybody can send us an encrypted message. Observe that we are the only ones who can decrypt it. Indeed, in order to pass from C to P we need to know d, and to compute d we need to know $\phi(n) = (p-1)(q-1)$. Actually, to know $\phi(n)$ is 'equivalent' to knowing p and q, as we may observe by writing

$$\begin{cases} p + q = n + 1 - \phi(n), \\ pq = n. \end{cases}$$

In other words, discovering $\phi(n)$ (and d) is as difficult as factoring $n = pq$. If p and q are large enough, this may take too much time. The security of RSA depends on the above (unproved) assumption.

Let us now see a practical application of the RSA encryption system through an example. Let us transform each letter (and space) as follows:

$$A = 01, \ B = 02, \ C = 03, \ D = 04, \ E = 05, \ F = 06, \ G = 07, \ H = 08,$$
$$I = 09, \ J = 10, \ K = 11, \ L = 12, \ M = 13, \ N = 14, \ O = 15, \ P = 16,$$
$$Q = 17, \ R = 18, \ S = 19, \ T = 20, \ U = 21, \ V = 22, \ W = 23, \ X = 24,$$
$$Y = 25, \ Z = 26, \ \text{space} = 00 \ .$$

Then the plaintext

$$\text{TOO MUCH TIME} \qquad\qquad (3.27)$$

becomes

$$2015150013210308002009 1305 \qquad\qquad (3.28)$$

We choose $n = 2867 = 47 \cdot 61$ (47 and 61 are prime numbers) and split the number in (3.28) into blocks not larger than 2867:

2015	1500	1321	0308	0020	0913	0500	
P_1	P_2	P_3	P_4	P_5	P_6	P_7	(3.29)

(we have added a couple of zeros at the end to have only 4-digit blocks). We have

$$\phi(2867) = 46 \cdot 60 = 2760 \ .$$

Since $2760 = 2^3 \cdot 3 \cdot 5 \cdot 23$, the number $h = 7$ satisfies $(h, \phi(2867)) = 1$. We now turn every block P_j in (3.29) into $C_j \equiv P_j^h \pmod{2867}$, with $0 \le C_j < 2867$. By doing this we obtain

$$P_1^h = 2015^7 \equiv 372 \pmod{2867}$$
$$P_2^h = 1500^7 \equiv 1147 \pmod{2867}$$
$$P_3^h = 1321^7 \equiv 387 \pmod{2867}$$
$$P_4^h = 308^7 \equiv 2736 \pmod{2867}$$
$$P_5^h = 20^7 \equiv 2047 \pmod{2867}$$
$$P_6^h = 913^7 \equiv 543 \pmod{2867}$$
$$P_7^h = 500^7 \equiv 1689 \pmod{2867}$$

so the cyphertext is

0372	1147	0387	2736	2047	0543	1689	
C_1	C_2	C_3	C_4	C_5	C_6	C_7	(3.30)

Let us recall that everybody knows $n = 2867$ and $h = 7$, but in order to turn

(3.30) back to (3.29) (and then to (3.27)) we need to know d (see (3.25)), that is, the solution of the congruence $dh \equiv 1 \pmod{\phi(n)}$, i.e.

$$7d \equiv 1 \pmod{2760} .$$

Strangers do not know $\phi(n) = \phi(2867) = 2760$, which comes from the decomposition of $n = 2867$, but we do. Therefore we know $d = 1183$, and we can go back to the plaintext:

$$372^{1183} \equiv 2015 \pmod{2867}$$
$$1147^{1183} \equiv 1500 \pmod{2867}$$
$$387^{1183} \equiv 1321 \pmod{2867}$$
$$2736^{1183} \equiv 308 \pmod{2867}$$
$$2047^{1183} \equiv 20 \pmod{2867}$$
$$543^{1183} \equiv 913 \pmod{2867}$$
$$1689^{1183} \equiv 500 \pmod{2867} .$$

The RSA encryption can also be used for digital signatures. If Bob needs to prove his identity to Alice, he simply needs to show her that he knows d, without revealing it. It works as follows. Bob chooses a message, for example, a part of a famous poem (assume for simplicity that this message can be written as a number $M \leq n$). Then he sends $K = M^d$ to Alice, who does not know d, but can compute $K^h = M^{dh} \equiv M \pmod{n}$, see the poem and be sure that Bob was the only one able to write K.

The literature on cryptography and number theory is very wide. See, for example, [78, 107, 145, 157, 167, 168, 171].

Exercises

1) Let $n = a_0 + 10a_1 + 10^2 a_2 + 10^3 a_3 + \ldots + 10^h a_h$ be a positive integer written in base 10. Prove that $n \equiv 0 \pmod{11}$ if and only if

$$a_0 - a_1 + a_2 - a_3 + \ldots + (-1)^h a_h \equiv 0 \pmod{11} .$$

2) Two lighthouses become visible now. Both lights switch on and off at regular intervals of time. The first one is visible for 7 seconds and off for 16 seconds, whereas the second one is visible for 8 seconds and off for 23 seconds. In how many seconds from now will both lights disappear from view together?

3) Prove that for no positive integer n do we have $2^n \equiv -1 \pmod{7}$.

4) Let $p \neq 3$ be a prime number. Prove that $p^2 + 2$ is not a prime number.

5) Let a be a positive integer and let p, q be different prime numbers such that $a^p \equiv a \pmod{q}$ and $a^q \equiv a \pmod{p}$. Prove that $a^{pq} \equiv a \pmod{pq}$.

6) Prove that for no integer $n > 1$ do we have $n \mid (2^n - 1)$.

7) Let $p \equiv 1 \pmod 4$ be a prime number. Use Wilson's theorem to show that $p \mid (s^2 + 1)$ for some $s \in \mathbb{N}$.

8) Assume that the positive integer n is not a square and let a satisfy $(a, n) = 1$. Prove the existence of integers x and y such that

$$0 \leq x < \sqrt{n}, \quad 0 \leq |y| < \sqrt{n}, \quad ax \equiv y \pmod n .$$

9) Let $p \in \mathcal{P}$. Prove that $a^p \equiv b^p \pmod p$ implies $a^p \equiv b^p \pmod{p^2}$.

10) Give a different proof of the infinitude of odd composite numbers which satisfy (3.12) by showing that if n is such a number, then the same is true for $2^n - 1$.

11) Where do we need $p > 3$ in the proof of Theorem 3.14?

4

Quadratic reciprocity and Fourier series

This chapter is devoted to the quadratic reciprocity law, proved by Gauss in 1796, when he was only 19 years old. See [118] for a story of the law. We will give two proofs. The second one will depend on Fourier series, and we are therefore going to start the parallel short course in Fourier analysis we talked about in the Introduction.

We now describe a general result on *polynomial congruences*.

Let $f(x) = a_n x^n + \ldots + a_0$ be a polynomial with integral coefficients, and let p be a prime number. Observe that the solutions of

$$f(x) \equiv 0 \pmod{p} \tag{4.1}$$

are residue classes (mod p). Lagrange proved the following result in 1770.

Theorem 4.1 *Let $f(x) = a_n x^n + a_{n-1} x^{n-1} + \ldots + a_1 x + a_0$ be a polynomial with integral coefficients. Let $p \in \mathcal{P}$ and assume that $p \nmid a_n$. Then the congruence*

$$f(x) \equiv 0 \pmod{p} \tag{4.2}$$

has at most n solutions (pairwise non-congruent (mod p)).

Proof The case $n = 0$ is not interesting and the case $n = 1$ follows from Theorem 3.5. We work by induction and assume that the result is true for polynomials of degree $n - 1$. Let $f(x)$ have degree n. If (4.2) has no solutions, then the result is true. If x_0 is a solution, then we write

$$f(x) = (x - x_0) f_1(x) + r$$

with $f_1(x)$ of degree $n - 1$ and $r \equiv 0 \pmod{p}$. Assume that $x_1 \not\equiv x_0 \pmod{p}$ is another solution of (4.2). Then

$$0 \equiv f(x_1) \equiv (x_1 - x_0) f_1(x_1) \pmod{p} .$$

65

Since p is a prime number and $x_1 - x_0 \not\equiv 0 \pmod{p}$ we have $f_1(x_1) \equiv 0 \pmod{p}$. Then the solutions of (4.2) which are not congruent to $x_0 \pmod{p}$ must be solutions of $f_1(x) \equiv 0 \pmod{p}$, of which, by the induction assumption, there are at most $n - 1$ (incongruent $\pmod{p}$). Hence (4.2) has at most n solutions. $\qquad\square$

Remark 4.2 If p is not a prime number, then the above theorem may fail. Indeed, the equation

$$x^2 - 1 \equiv 0 \pmod{8}$$

has four solutions: $x = \pm 1, \pm 3$.

Remark 4.3 The assumption $p \nmid a_n$ enables us to see the 'true' degree of a polynomial inside a congruence $\pmod{p}$. As an example, the polynomial $15x^7 + x^6 - 3x + 2$ has degree 7, but the congruence $15x^7 + x^6 - 3x + 2 \equiv 0 \pmod{5}$ should be rewritten as $x^6 - 3x + 2 \equiv 0 \pmod{5}$.

4.1 Quadratic residues

We consider the *quadratic congruence*

$$ax^2 + bx + c \equiv 0 \pmod{p}, \tag{4.3}$$

where $a, b, c \in \mathbb{Z}$ and $\mathcal{P} \ni p \nmid a$. The case $p = 2$ is rather simple, so we shall assume that p is odd. Since a is invertible $\pmod{p}$, we may rewrite (4.3) as

$$x^2 + \beta x + \gamma \equiv 0 \pmod{p}.$$

Moreover we can assume that β is even, since we can always replace βx by $(\beta + p)x$. We therefore investigate

$$x^2 + 2\delta x + \gamma \equiv 0 \pmod{p}, \qquad (x + \delta)^2 \equiv \delta^2 - \gamma \pmod{p}.$$

That is, we consider congruences of the form

$$z^2 \equiv a \pmod{p}, \tag{4.4}$$

where $\mathcal{P} \ni p \nmid a$.

Definition 4.4 If (4.4) admits a solution, then we say that a is a *quadratic residue* $\pmod{p}$. Otherwise we say that a is a *quadratic non-residue* $\pmod{p}$.

Theorem 4.5 *Let p be an odd prime number. Then there are exactly $(p-1)/2$ quadratic residues (mod p):*

$$1^2, 2^2, \ldots, \left(\frac{p-1}{2}\right)^2 . \tag{4.5}$$

Proof Of course (4.4) has a solution when a equals one of the numbers in (4.5). Moreover, assume that $r^2 \equiv s^2 \pmod{p}$ with $1 \leq s \leq r \leq \frac{p-1}{2}$. Then

$$r^2 - s^2 = (r - s)(r + s) \equiv 0 \pmod{p} .$$

Since $2 \leq r + s \leq p - 1$, we have $r \equiv s \pmod{p}$. Now assume that a is a quadratic residue with $a \equiv t^2 \pmod{p}$ and $\frac{p+1}{2} < t \leq p - 1$. We claim that $a \equiv u^2 \pmod{p}$ with $1 \leq u \leq \frac{p-1}{2}$. Indeed, $t^2 \equiv (t - p)^2 \pmod{p}$ and $-\frac{p-1}{2} \leq t - p \leq -1$. Hence $t^2 \equiv u^2 \pmod{p}$ with $1 \leq u \leq \frac{p-1}{2}$. □

Now we define the Legendre symbol, introduced by Legendre in 1798. Let a be an integer and let p be an odd prime number. We write

$$\left(\frac{a}{p}\right)_L := \begin{cases} 1 & \text{if } p \nmid a \text{ and } a \text{ is a quadratic residue } \pmod{p}, \\ -1 & \text{if } p \nmid a \text{ and } a \text{ is a quadratic non-residue } \pmod{p}, \\ 0 & \text{if } p \mid a. \end{cases}$$

Observe that $\left(\frac{1}{p}\right)_L = 1$ for every p, because the equation $x^2 \equiv 1 \pmod{p}$ has the solution $x = 1$.

Theorem 4.6 (Euler's criterion) *Let p be an odd prime number. Then for every integer a we have*

$$\left(\frac{a}{p}\right)_L \equiv a^{(p-1)/2} \pmod{p} . \tag{4.6}$$

Proof If $p \mid a$, then $0 \equiv a^{(p-1)/2} \pmod{p}$ and the theorem is proved. Assume that $p \nmid a$. If $\left(\frac{a}{p}\right)_L = 1$, then there exists z such that $z^2 \equiv a \pmod{p}$, and $p \nmid z$. Then Fermat's little theorem implies

$$a^{(p-1)/2} \equiv z^{p-1} \equiv 1 \equiv \left(\frac{a}{p}\right)_L \pmod{p} .$$

Now assume that $\left(\frac{a}{p}\right)_L = -1$, so that the equation $z^2 \equiv a \pmod{p}$ has no solutions. Observe that for every $j = 1, \ldots, p - 1$, Theorem 3.5 shows the existence of a unique $1 \leq \ell_j \leq p - 1$ such that $j\ell_j \equiv a \pmod{p}$. For every j we have $j \neq \ell_j$ (because the equation $z^2 \equiv a \pmod{p}$ has no solutions). In this way we have subdivided the set $\{1, 2, \ldots, p - 1\}$ into $(p - 1)/2$ pairs $\left(j, \ell_j\right)$.

Then, by Wilson's theorem,

$$-1 \equiv (p-1)! \equiv \prod_{j=1}^{p-1} j \equiv \prod_{j=1}^{(p-1)/2} \left(j\ell_j \right) \equiv a^{(p-1)/2} \pmod{p} \ .$$

□

Remark 4.7   In particular, if p is an odd prime number, then the equation $x^2 \equiv -1 \pmod{p}$ has a solution if and only if $p \equiv 1 \pmod{4}$.

Theorem 4.8   *Let p be an odd prime number. Then for every $a, b \in \mathbb{Z}$ we have*

$$\left(\frac{ab}{p} \right)_L = \left(\frac{a}{p} \right)_L \left(\frac{b}{p} \right)_L \ . \tag{4.7}$$

Proof   By Euler's criterion we have

$$\left(\frac{ab}{p} \right)_L \equiv (ab)^{(p-1)/2} \equiv a^{(p-1)/2} b^{(p-1)/2} \equiv \left(\frac{a}{p} \right)_L \left(\frac{b}{p} \right)_L \pmod{p} \ . \tag{4.8}$$

In order to end the proof we observe that

$$\left| \left(\frac{ab}{p} \right)_L - \left(\frac{a}{p} \right)_L \left(\frac{b}{p} \right)_L \right| \leq 2 \ .$$

Then (4.8) implies (4.7).

□

Theorem 4.9 (Gauss' lemma)   *Let p be an odd prime number and let a be an integer such that $p \nmid a$. Let*

$$m = \mathrm{card} \left\{ x \in \mathbb{N} : 1 \leq x < \frac{p}{2} \ \text{and} \ \frac{p}{2} < ax - p\left[\frac{ax}{p} \right] < p \right\} , \tag{4.9}$$

where $[\alpha]$ is the largest integer $\leq \alpha$. Then

$$\left(\frac{a}{p} \right)_L = (-1)^m \ . \tag{4.10}$$

Observe that $ax - p\left[\frac{ax}{p} \right]$ is the residue of the division of ax by p.

Proof   Let $t_1, t_2, \ldots, t_m$ be the residues $r_{(x)} = ax - p\left[\frac{ax}{p} \right]$ which are larger than $p/2$. Let $s_1, s_2, \ldots, s_{\frac{p-1}{2}-m}$ be the other residues. Observe that $x \neq y$ implies $r_{(x)} \neq r_{(y)}$, for otherwise we would have $p \mid a(x-y)$, hence $x = y$. We have $0 < p - t_j < p/2$ for every $j = 1, \ldots, m$. We claim that $p - t_j \neq s_h$ for every $j = 1, \ldots, m$ and $h = 1, \ldots, \frac{p-1}{2} - m$. Otherwise (if $t_j = r_{(x)}$ and $s_h = r_{(y)}$) we would have $-t_j \equiv s_h \pmod{p}$ and then $-ax \equiv ay \pmod{p}$, which is impossible because $0 < x + y < p$ and therefore $p \nmid a(x+y)$. Then

$$\left\{ 1, 2, \ldots, \frac{p-1}{2} \right\} \tag{4.11}$$

$$= \left\{ p - t_j : j = 1, \ldots, m \right\} \cup \left\{ s_h : h = 1, \ldots, \frac{p-1}{2} - m \right\} .$$

Hence

$$\left(\frac{p-1}{2} \right)! = \prod_{j=1}^{m} \left(p - t_j \right) \prod_{h=1}^{\frac{p-1}{2}-m} s_h \equiv (-1)^m \prod_{j=1}^{m} t_j \prod_{h=1}^{\frac{p-1}{2}-m} s_h \equiv (-1)^m \prod_{x=1}^{(p-1)/2} ax$$

$$\equiv (-1)^m a^{(p-1)/2} \left(\frac{p-1}{2} \right)! \pmod{p} .$$

Since $p \nmid \left(\frac{p-1}{2} \right)!$, Remark 3.2 implies

$$(-1)^m a^{(p-1)/2} \equiv 1 \pmod{p} .$$

Finally, by Euler's criterion,

$$(-1)^m \equiv a^{(p-1)/2} \equiv \left(\frac{a}{p} \right)_L \pmod{p} .$$

This implies (4.10). □

We need one more lemma.

Lemma 4.10 *Let p be an odd prime number and let q be a different prime number. If $q \neq 2$, we have*

$$\left(\frac{q}{p} \right)_L = (-1)^{\sum_{k=1}^{(p-1)/2} [qk/p]} , \tag{4.12}$$

while for $q = 2$,

$$\left(\frac{2}{p} \right)_L = (-1)^{(p^2-1)/8} . \tag{4.13}$$

Proof Let m be as in (4.9) with $a = q$. Among the residues on division of qk by p, let $t_1, \ldots, t_m$ denote those larger than $p/2$ and let $s_1, \ldots, s_{\frac{p-1}{2}-m}$ denote those smaller than $p/2$. Then

$$q \frac{p^2 - 1}{8} = q \sum_{k=1}^{\frac{p-1}{2}} k = p \sum_{k=1}^{\frac{p-1}{2}} \frac{qk}{p} = p \sum_{k=1}^{\frac{p-1}{2}} \left[\frac{qk}{p} \right] + \sum_{j=1}^{m} t_j + \sum_{j=1}^{\frac{p-1}{2}-m} s_j . \tag{4.14}$$

By (4.11) we have

$$\frac{p^2 - 1}{8} = \sum_{k=1}^{\frac{p-1}{2}} k = \sum_{j=1}^{m} \left(p - t_j \right) + \sum_{j=1}^{\frac{p-1}{2}-m} s_j . \tag{4.15}$$

Subtracting (4.15) from (4.14) we obtain

$$(q-1)\frac{p^2-1}{8} = p\sum_{k=1}^{\frac{p-1}{2}}\left[\frac{qk}{p}\right] - mp + 2\sum_{j=1}^{m}t_j . \qquad (4.16)$$

Assume that q is odd. Then

$$0 \equiv (q-1)\frac{p^2-1}{8} \equiv p\sum_{k=1}^{\frac{p-1}{2}}\left[\frac{qk}{p}\right] - mp \equiv \sum_{k=1}^{\frac{p-1}{2}}\left[\frac{qk}{p}\right] - m \pmod 2 ,$$

that is,

$$\sum_{k=1}^{\frac{p-1}{2}}\left[\frac{qk}{p}\right] \equiv m \pmod 2 .$$

By Theorem 4.9 this implies (4.12). We now prove (4.13). Writing $q = 2$ in
(4.16) we obtain (since $\left[\frac{2k}{p}\right] = 0$ when $1 \le k \le \frac{p-1}{2}$)

$$\frac{p^2-1}{8} = -mp + 2\sum_{j=1}^{m}t_j .$$

Since p is odd we deduce that

$$\frac{p^2-1}{8} \equiv m \pmod 2 .$$

This completes the proof. $\square$

We are now in a position to state and prove the celebrated Gauss *quadratic
reciprocity law*, which relates in a very simple way the solubility of the congruences

$$x^2 \equiv q \pmod p , \qquad x^2 \equiv p \pmod q .$$

Theorem 4.11 (Quadratic reciprocity law)   *Let $p \ne q$ be odd prime numbers.
Then*

$$\left(\frac{q}{p}\right)_L \left(\frac{p}{q}\right)_L = (-1)^{\left(\frac{p-1}{2}\right)\left(\frac{q-1}{2}\right)} .$$

Proof   Let

$$S = \left\{(k,\ell) \in \mathbb{N}^2 : k \le \frac{p-1}{2} \text{ and } \ell \le \frac{q-1}{2}\right\},$$

$$S_1 = \{(k,\ell) \in S : \ell < qk/p\},$$

$$S_2 = \{(k,\ell) \in S : k < p\ell/q\}.$$

Observe that for no choice of k and ℓ can we have $qk = p\ell$. Then $S = S_1 \cup S_2$.

In the figure below, S_1 and S_2 consist of the integer points inside the two triangles (we have observed that no integer point belongs to the common side of the two triangles).

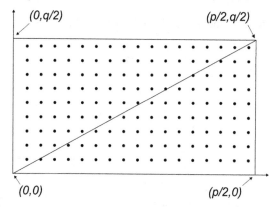

We have

$$\operatorname{card} S_1 = \sum_{k=1}^{\frac{p-1}{2}} \left[\frac{qk}{p} \right], \qquad \operatorname{card} S_2 = \sum_{\ell=1}^{\frac{q-1}{2}} \left[\frac{p\ell}{q} \right].$$

Then

$$\left(\frac{p-1}{2} \right) \left(\frac{q-1}{2} \right) = \operatorname{card} S = \operatorname{card} S_1 + \operatorname{card} S_2 = \left(\sum_{k=1}^{\frac{p-1}{2}} \left[\frac{qk}{p} \right] \right) + \left(\sum_{\ell=1}^{\frac{q-1}{2}} \left[\frac{p\ell}{q} \right] \right).$$

This identity and the previous lemma complete the proof. $\square$

We now work out two examples.

We study the solubility of the congruence

$$x^2 \equiv 8783 \pmod{15671} \tag{4.17}$$

(8783 and 15671 are prime numbers). By the quadratic reciprocity law we have

$$\left(\frac{8783}{15671} \right)_L = (-1)^{\left(\frac{8783-1}{2} \right) \left(\frac{15671-1}{2} \right)} \left(\frac{15671}{8783} \right)_L = -\left(\frac{15671}{8783} \right)_L = -\left(\frac{6888}{8783} \right)_L,$$

because $x^2 \equiv q \pmod{p}$ is equivalent to $x^2 \equiv q + \ell p \pmod{p}$. Since $6888 = 2^3 \times 3 \times 7 \times 41$, Theorem 4.8 implies

$$-\left(\frac{6888}{8783} \right)_L = -\left(\frac{2}{8783} \right)_L^3 \left(\frac{3}{8783} \right)_L \left(\frac{7}{8783} \right)_L \left(\frac{41}{8783} \right)_L$$

$$= -\left(\frac{2}{8783} \right)_L \left(\frac{3}{8783} \right)_L \left(\frac{7}{8783} \right)_L \left(\frac{41}{8783} \right)_L.$$

We compute each of the above terms. By (4.13) we obtain

$$\left(\frac{2}{8783}\right)_L = (-1)^{\frac{1}{8}(8783^2-1)} = (-1)^{9642636} = 1 \ .$$

Since $8783 \equiv 2 \pmod 3$, the quadratic reciprocity law implies

$$\left(\frac{3}{8783}\right)_L = (-1)^{(\frac{3-1}{2})(\frac{8783-1}{2})}\left(\frac{8783}{3}\right)_L = -\left(\frac{8783}{3}\right)_L = -\left(\frac{2}{3}\right)_L = -(-1)^{\frac{3^2-1}{8}} = 1 \ ,$$

$$\left(\frac{7}{8783}\right)_L = (-1)^{(\frac{7-1}{2})(\frac{8783-1}{2})}\left(\frac{8783}{7}\right)_L = -\left(\frac{8783}{7}\right)_L = -\left(\frac{5}{7}\right)_L$$

$$= -(-1)^{(\frac{5-1}{2})(\frac{7-1}{2})}\left(\frac{7}{5}\right)_L = -\left(\frac{7}{5}\right)_L = -\left(\frac{2}{5}\right)_L = -(-1)^{\frac{5^2-1}{8}} = 1 \ ,$$

$$\left(\frac{41}{8783}\right)_L = (-1)^{(\frac{41-1}{2})(\frac{8783-1}{2})}\left(\frac{8783}{41}\right)_L = \left(\frac{8783}{41}\right)_L = \left(\frac{9}{41}\right)_L = \left(\frac{3}{41}\right)_L^2 = 1$$

by Theorem 4.8. Hence

$$\left(\frac{8783}{15671}\right)_L = -1$$

and the equation (4.17) has no solutions. Note that we could have used (4.6), for example, $\left(\frac{5}{7}\right)_L \equiv 5^3 \equiv -1 \pmod 7$ and then $\left(\frac{5}{7}\right)_L = -1$.

As a second example we consider the congruence

$$x^2 \equiv -42 \pmod{61} \ .$$

Since 61 is a prime number, Theorem 4.8 implies

$$\left(\frac{-42}{61}\right)_L = \left(\frac{-1}{61}\right)_L \left(\frac{2}{61}\right)_L \left(\frac{3}{61}\right)_L \left(\frac{7}{61}\right)_L \ .$$

By (4.6) we have

$$\left(\frac{-1}{61}\right)_L \equiv (-1)^{(61-1)/2} \pmod{61} \ ,$$

that is,

$$\left(\frac{-1}{61}\right)_L = 1 \ .$$

By (4.13) we have

$$\left(\frac{2}{61}\right)_L = (-1)^{(61^2-1)/8} = -1 \ .$$

We have

$$\left(\frac{3}{61}\right)_L = \left(\frac{61}{3}\right)_L (-1)^{\left(\frac{61-1}{2}\right)\left(\frac{3-1}{2}\right)} = \left(\frac{61}{3}\right)_L = \left(\frac{1}{3}\right)_L = 1 .$$

Finally

$$\left(\frac{7}{61}\right)_L = \left(\frac{61}{7}\right)_L (-1)^{\left(\frac{61-1}{2}\right)\left(\frac{7-1}{2}\right)} = \left(\frac{61}{7}\right)_L = \left(\frac{5}{7}\right)_L$$

$$= \left(\frac{7}{5}\right)_L (-1)^{\left(\frac{7-1}{2}\right)\left(\frac{5-1}{2}\right)} = \left(\frac{7}{5}\right)_L = \left(\frac{2}{5}\right)_L = (-1)^{\frac{5^2-1}{8}} = -1 .$$

Hence

$$\left(\frac{-42}{61}\right)_L = 1$$

and the congruence $x^2 \equiv -42 \pmod{61}$ has solutions (indeed 18 and 43 are solutions, because $18^2 = 324 = 61 \cdot 6 - 42$ and $43^2 = 1849 = 61 \cdot 31 - 42$). Another way to handle this example is as follows. Note that

$$\left(\frac{-42}{61}\right)_L = \left(\frac{19}{61}\right)_L .$$

Quadratic reciprocity converts this problem to evaluating

$$\left(\frac{19}{61}\right)_L = \left(\frac{4}{19}\right)_L = \left(\frac{2}{19}\right)_L \left(\frac{2}{19}\right)_L = 1 .$$

We end this section by computing $\left(\frac{3}{p}\right)_L$ for every prime number $p > 3$. Arguing as before, we obtain

$$\left(\frac{3}{p}\right)_L = \left(\frac{p}{3}\right)_L (-1)^{\left(\frac{p-1}{2}\right)\left(\frac{3-1}{2}\right)} = \left(\frac{p}{3}\right)_L (-1)^{\left(\frac{p-1}{2}\right)} .$$

We recall (4.6). If $p \equiv 1 \pmod 3$ we have

$$\left(\frac{p}{3}\right)_L = \left(\frac{1}{3}\right)_L = 1 ,$$

while if $p \equiv 2 \pmod 3$ we have

$$\left(\frac{p}{3}\right)_L = \left(\frac{-1}{3}\right)_L \equiv (-1)^{\frac{3-1}{2}} \equiv -1 \pmod p .$$

Moreover,

$$(-1)^{\left(\frac{p-1}{2}\right)} = \begin{cases} +1 & \text{if } p \equiv 1 \pmod 4 , \\ -1 & \text{if } p \equiv 3 \pmod 4 . \end{cases}$$

Then $\left(\frac{3}{p}\right)_L = 1$ if and only if

$$p \equiv 1 \pmod 3 \text{ and } p \equiv 1 \pmod 4$$

or

$$p \equiv 2 \pmod 3 \text{ and } p \equiv 3 \pmod 4 .$$

Then $\left(\frac{3}{p}\right)_L = 1$ if and only if $p \equiv \pm 1 \pmod{12}$, while there are no solutions for $p \equiv \pm 5 \pmod{12}$. Note that any odd prime $p > 3$ must satisfy $p \equiv \pm 1 \pmod{12}$ or $p \equiv \pm 5 \pmod{12}$.

4.2 Gauss sums

We shall give another proof of Theorem 4.11, based on Gauss sums and Fourier series. In this section we introduce Gauss sums.

Definition 4.12 Let $0 \neq a \in \mathbb{Z}$ and $q \in \mathbb{N}$. Assume that $(a, q) = 1$. We define the *Gauss sum*

$$S(q, a) := \sum_{n=1}^{q} e^{2\pi i a n^2 / q} . \tag{4.18}$$

Lemma 4.13 Let $0 \neq a \in \mathbb{Z}$ and $q_1, q_2 \in \mathbb{N}$ such that $(q_1, q_2) = 1$ and $(a, q_1 q_2) = 1$. Then

$$S(q_1 q_2, a) = S(q_1, q_2 a) \, S(q_2, q_1 a) .$$

Proof We recall Theorem 3.4, which says that if $R = \{x_j\}_{j=1}^{q_1}$ and $S = \{y_k\}_{k=1}^{q_2}$ are complete sets of residues mod q_1 and mod q_2, respectively, then the set $\{q_2 x_j + q_1 y_k\}_{1 \leq j \leq q_1, \, 1 \leq k \leq q_2}$ is a complete set of residues mod $q_1 q_2$. Then

$$S(q_1 q_2, a) = \sum_{n=1}^{q_1 q_2} e^{2\pi i a n^2 /(q_1 q_2)} = \sum_{j=1}^{q_1} \sum_{k=1}^{q_2} e^{2\pi i a \left(q_2 x_j + q_1 y_k\right)^2 /(q_1 q_2)} .$$

Since $\left(q_2 x_j + q_1 y_k\right)^2 \equiv q_2^2 x_j^2 + q_1^2 y_k^2 \pmod{q_1 q_2}$ we deduce that

$$S(q_1 q_2, a) = \sum_{j=1}^{q_1} e^{2\pi i a q_2 x_j^2 / q_1} \sum_{k=1}^{q_2} e^{2\pi i a q_1 y_k^2 / q_2} = S(q_1, q_2 a) S(q_2, q_1 a) .$$

$\square$

Gauss sums and the Legendre symbol are related by the following result.

Lemma 4.14 *Let $0 \neq a \in \mathbb{Z}$ and let p be an odd prime number such that $(a, p) = 1$. Then*

$$S(p,a) = \left(\frac{a}{p}\right)_L S(p,1) .$$

Proof In the proof of Theorem 4.5 we have seen that the equation

$$x^2 \equiv m \pmod{p} \tag{4.19}$$

has one solution if $m \equiv 0 \pmod{p}$, otherwise it has zero or two solutions. Then the number of solutions of (4.19) equals $1 + \left(\frac{m}{p}\right)_L$ and we can write

$$S(p,a) = \sum_{n=1}^{p} e^{2\pi i (an^2/p)} = \sum_{m=1}^{p} \left(1 + \left(\frac{m}{p}\right)_L\right) e^{2\pi i am/p} \tag{4.20}$$

$$= \sum_{m=1}^{p} e^{2\pi i am/p} + \sum_{m=1}^{p} \left(\frac{m}{p}\right)_L e^{2\pi i am/p} = \sum_{m=1}^{p} \left(\frac{m}{p}\right)_L e^{2\pi i am/p} .$$

Indeed, by Theorem 3.3 and the assumption $(a, p) = 1$ we deduce that $\{am\}_{m=1}^{p}$ is a complete set of residues mod p. Then

$$\sum_{m=1}^{p} e^{2\pi i am/p} = \sum_{h=1}^{p} e^{2\pi i h/p} = 0$$

(observe that $\sum_{h=1}^{p} e^{2\pi i h/p}$ is the sum in $\mathbb{C}$ of the pth roots of unity). Since $(a, p) = 1$, Corollary 3.6 implies the existence of an inverse a^* of a. Then (4.20) and Theorem 4.8 give

$$S(p,a) = \sum_{h=1}^{p} \left(\frac{a^* h}{p}\right)_L e^{2\pi i h/p} = \left(\frac{a^*}{p}\right)_L \sum_{h=1}^{p} \left(\frac{h}{p}\right)_L e^{2\pi i h/p} = \left(\frac{a}{p}\right)_L \sum_{h=1}^{p} \left(\frac{h}{p}\right)_L e^{2\pi i h/p} ,$$

since $1 = \left(\frac{1}{p}\right)_L = \left(\frac{a}{p}\right)\left(\frac{a^*}{p}\right)$. In particular we have

$$S(p,1) = \sum_{h=1}^{p} \left(\frac{h}{p}\right)_L e^{2\pi i h/p} .$$

Then

$$S(p,a) = \left(\frac{a}{p}\right)_L S(p,1) .$$

$\square$

4.3 Fourier series

In this section we start to discuss Fourier analysis. See [81, 159, 161, 164, 185]. For the real analysis results see [82, 162, 182].

A complex vector space $\mathcal{H}$ with inner product $\langle \cdot, \cdot \rangle$ and norm $\|f\| := \sqrt{\langle f, f \rangle}$ is a (separable) Hilbert space if it is complete and separable with respect to the distance $d(f, g) := \|f - g\|$. A sequence $\{u_n\} \subset \mathcal{H}$ is *orthonormal* if $\langle u_n, u_m \rangle = \delta_{n,m}$ for every m, n (here $\delta_{n,m} = 0$ if $n \neq m$, while $\delta_{n,n} = 1$). We say that $\{u_n\}_{n=1}^{+\infty}$ is an *orthonormal basis* if it satisfies one of the following equivalent conditions.

Theorem 4.15 *Let $\mathcal{H}$ be a separable Hilbert space and let $\{u_n\}_{n=1}^{+\infty}$ be a sequence in $\mathcal{H}$. The following conditions are equivalent.*

(i) *The finite linear combinations of the terms u_n are dense in $\mathcal{H}$.*
(ii) *Let $f \in \mathcal{H}$. If $\langle f, u_n \rangle = 0$ for every n, then $f = 0$.*
(iii) *For every $f \in \mathcal{H}$ we have $\left\| f - \sum_{n=1}^{N} \langle f, u_n \rangle u_n \right\| \to 0$ as $N \to +\infty$.*
(iv) *For every $f, g \in \mathcal{H}$ we have $\langle f, g \rangle = \sum_n \langle f, u_n \rangle \overline{\langle g, u_n \rangle}$.*
(v) *For every $f \in \mathcal{H}$ we have $\|f\|^2 = \sum_n |\langle f, u_n \rangle|^2$ (Parseval identity).*

Proof We assume (i). Let $f \in \mathcal{H}$, then there is a sequence $\{g_m\}_{m=1}^{+\infty}$ of finite linear combinations of the terms u_n, such that $\|g_m - f\| \to 0$ as $m \to +\infty$. Assume that $\langle f, u_n \rangle = 0$ for every n. Then $\langle f, g_m \rangle = 0$ for every m. By the Cauchy–Schwartz inequality we have

$$\|f\|^2 = \langle f, f \rangle = \langle f, f - g_m \rangle \leq \|f\| \, \|f - g_m\| \longrightarrow 0$$

as $m \to +\infty$. Then (i) implies (ii). Now we assume (ii). For every positive integer N we define the partial sum

$$S_N f := \sum_{n=1}^{N} \langle f, u_n \rangle u_n$$

and observe that $\langle f - S_N f, S_N f \rangle = 0$. Then[1]

$$\|f\|^2 = \langle f - S_N f + S_N f, f - S_N f + S_N f \rangle \tag{4.21}$$

$$= \|f - S_N f\|^2 + \langle S_N f, S_N f \rangle = \|f - S_N f\|^2 + \sum_{m,n=1}^{N} \langle \langle f, u_m \rangle u_m, \langle f, u_n \rangle u_n \rangle$$

$$= \|f - S_N f\|^2 + \sum_{n=1}^{N} |\langle f, u_n \rangle|^2 \; .$$

---

[1] We recall that $\langle ax + by, z \rangle = a \langle x, z \rangle + b \langle y, z \rangle$, and that $\langle x, y \rangle = \overline{\langle y, x \rangle}$. Hence $\langle x, ay + bz \rangle = \overline{a} \langle x, y \rangle + \overline{b} \langle x, z \rangle$.

Letting $N \to +\infty$ we obtain *Bessel's inequality*

$$\sum_n |\langle f, u_n \rangle|^2 \leq \|f\|^2 , \tag{4.22}$$

which implies the convergence of the series $\sum_n |\langle f, u_n \rangle|^2$. Then $S_N f$ is a Cauchy sequence in $\mathcal{H}$ and therefore it converges to an element $g \in \mathcal{H}$. For every j we have $\langle f - S_N f, u_j \rangle = 0$ when N is large enough, and then $\langle f - g, u_j \rangle = 0$. Hence $f = g$ by (ii). Then $f = \sum_n \langle f, u_n \rangle u_n$ (in norm). We now assume (iii). Arguing as in (4.21) we obtain

$$\langle f, g \rangle = \langle f - S_N f, g - S_N g \rangle + \langle S_N f, S_N g \rangle .$$

By the Cauchy–Schwartz inequality and (iii) we have

$$\langle f - S_N f, g - S_N g \rangle \leq \|f - S_N f\| \, \|g - S_N g\| \longrightarrow 0 .$$

Furthermore

$$\langle S_N f, S_N g \rangle = \sum_{m,n=1}^{N} \langle \langle f, u_m \rangle u_m, \langle g, u_n \rangle u_n \rangle = \sum_{n=1}^{N} \langle f, u_n \rangle \overline{\langle g, u_n \rangle} .$$

Letting $N \to +\infty$ we obtain (iv). Observe that (v) is a particular case of (iv). Finally, let us assume (v). Again from (4.21) we see that $\|f - S_N f\| \to 0$. Since $S_N f$ is a finite linear combinations of the terms u_n, we obtain (i). $\quad\square$

We are interested in periodic functions and we assume that the period is equal to 1. Periodic functions can be seen as functions on the 1-dimensional torus $\mathbb{T} := \mathbb{R}/\mathbb{Z}$. In some problems this amounts to considering functions defined on $[0, 1)$ or on another interval of length 1. For the sake of symmetry we shall often use the interval $[-1/2, 1/2)$. We point out that in some situations it is misleading to identify $\mathbb{T}$ with an interval of length 1. For example, a function can be continuous on $[-1/2, 1/2)$, but it can have a discontinuous periodic continuation.

Consider the Hilbert space $L^2(\mathbb{T})$ of square integrable functions on $\mathbb{T}$. Here

$$\langle f, g \rangle := \int_{\mathbb{T}} f(x) \overline{g(x)} \, dx \quad ; \quad \|f\|_{L^2(\mathbb{T})} = \left\{ \int_{\mathbb{T}} |f(x)|^2 \, dx \right\}^{1/2} .$$

We shall see that the exponentials $\{e_n\}_{n \in \mathbb{Z}} := \{e^{2\pi i n x}\}_{n \in \mathbb{Z}}$ are an orthonormal basis of $L^2(\mathbb{T})$. For every $f \in L^2(\mathbb{T})$ the projections

$$\widehat{f}(n) := \langle f, e_n \rangle = \int_{\mathbb{T}} f(x) \, e^{-2\pi i n x} \, dx$$

are called *Fourier coefficients* of f. Indeed, we shall prove (Corollary 6.14) the following result.

Theorem 4.16   *For every $f \in L^2(\mathbb{T})$ we have the Parseval identity*

$$\sum_{n=-\infty}^{+\infty} \left|\widehat{f}(n)\right|^2 = \int_{\mathbb{T}} |f(x)|^2 \, dx \, . \tag{4.23}$$

Hence

$$f(x) = \sum_{n=-\infty}^{+\infty} \widehat{f}(n) \, e^{2\pi i n x} \tag{4.24}$$

in the L^2 norm.

The RHS of (4.24) is called the *Fourier series* of $f(x)$.

We write $\ell^2(\mathbb{Z})$ for the Hilbert space of complex sequences $a = \{a_n\}_{n=-\infty}^{+\infty}$ which satisfy

$$\sum_{n=-\infty}^{+\infty} |a_n|^2 < +\infty \, .$$

If $a, b \in \ell^2(\mathbb{Z})$, the inner product and the norm, respectively, are

$$\langle a, b \rangle := \sum_{n=-\infty}^{+\infty} a_n \overline{b}_n \quad \text{and} \quad \|a\| = \left\{\sum_{n=-\infty}^{+\infty} |a_n|^2\right\}^{1/2} \, .$$

There is one and only one function $f \in L^2(\mathbb{T})$ such that $a_n = \widehat{f}(n)$. Moreover, $f \leftrightarrow \widehat{f}$ is a linear isometry between $L^2(\mathbb{T})$ and $\ell^2(\mathbb{Z})$. See, for example, [182].

The natural convergence of a Fourier series is in the L^2 sense, and the pointwise convergence for continuous functions may fail, see Theorem 4.19. We observe that if $\sum_{n\in\mathbb{Z}} \widehat{f}(n) \, e^{2\pi i n x}$ converges absolutely, that is, $\sum_{n\in\mathbb{Z}} |\widehat{f}(n)| < +\infty$, then $\sum_{n\in\mathbb{Z}} \widehat{f}(n) \, e^{2\pi i n x}$ converges uniformly to $f(x)$. Unfortunately the absolute convergence fails in many relevant cases. Then we look for other conditions which imply the pointwise convergence.

Definition 4.17   We say that a function f on an interval $[a, b]$ is *piecewise continuous* on $[a, b]$ if

(i) f is continuous everywhere on $[a, b]$ except at finitely many points $x_1, \ldots, x_k$.

(ii) $f\left(x_j^+\right) := \lim_{x \to x_j^+} f(x)$ and $f\left(x_j^-\right) := \lim_{x \to x_j^-} f(x)$ exist and are finite for every $j = 1, \ldots, k$ (if $x_j = a$ or $x_j = b$, then we require only one limit).

We say that a function f on an interval $[a, b]$ is *piecewise smooth* on $[a, b]$ if f' is a piecewise continuous function on $[a, b]$.

Theorem 4.18 *Let $f : \mathbb{R} \to \mathbb{C}$ be 1-periodic and piecewise smooth, say on $\left[-\frac{1}{2}, \frac{1}{2}\right)$. Then for every $x_0 \in \mathbb{R}$ we have, as $N \to +\infty$,*

$$S_N f(x_0) := \sum_{n=-N}^{+N} \widehat{f}(n)\, e^{2\pi i n x_0} \longrightarrow \frac{1}{2}\left(f\left(x_0^-\right) + f\left(x_0^+\right)\right).$$

Proof We call $S_N f$ the *Fourier partial sum* of f. Observe that if $\ell(x) = f(x + x_0)$, then $\widehat{\ell}(n) = \widehat{f}(n)\, e^{2\pi i n x_0}$. Then we may assume that $x_0 = 0$. Moreover, we can replace $f(x)$ with

$$k(x) = f(x) - \frac{f(0^+) + f(0^-)}{2}.$$

Then $\widehat{f}(n) = \widehat{k}(n)$ for every $n \neq 0$ and we may assume that $f(0^-) = -f(0^+)$. In this way the theorem reduces to the claim that

$$S_N f(0) \to 0.$$

Observe that the function $g(x) = \frac{1}{2}\{f(x) + f(-x)\}$ is piecewise smooth, continuous at $x = 0$ and satisfies $g(0) = 0$. Since $\sum_{n=-N}^{+N} e^{-2\pi i n x} = \sum_{n=-N}^{+N} e^{2\pi i n x}$, we obtain

$$S_N f(0) = \sum_{n=-N}^{+N} \widehat{f}(n) = \sum_{n=-N}^{+N} \int_{-1/2}^{1/2} f(x)\, e^{-2\pi i n x}\, dx$$

$$= \frac{1}{2}\left(\int_{-1/2}^{1/2} f(x) \sum_{n=-N}^{+N} e^{-2\pi i n x}\, dx + \int_{-1/2}^{1/2} f(-x) \sum_{n=-N}^{+N} e^{2\pi i n x}\, dx\right)$$

$$= \int_{-1/2}^{1/2} \frac{f(x) + f(-x)}{2} \sum_{n=-N}^{+N} e^{-2\pi i n x}\, dx$$

$$= \sum_{n=-N}^{+N} \int_{-1/2}^{1/2} \frac{f(x) + f(-x)}{2}\, e^{-2\pi i n x}\, dx = \sum_{n=-N}^{+N} \widehat{g}(n) = S_N g(0).$$

Now we prove that $S_N g(0) \to 0$. Indeed, let

$$h(x) := \frac{g(x)}{e^{2\pi i x} - 1}.$$

We observe that

$$h(x) \sim \frac{g(x)}{2\pi i x} \to \begin{cases} \frac{1}{2\pi i}\, g'(0^-) & \text{as } x \to 0^-, \\ \frac{1}{2\pi i}\, g'(0^+) & \text{as } x \to 0^+. \end{cases}$$

The function h is piecewise continuous and therefore bounded. Then Bessel's inequality (4.22) implies

$$\widehat{h}(n) \to 0$$

as $|n| \to +\infty$. Now observe that

$$\widehat{g}(n) = \int_{-1/2}^{1/2} g(x)\, e^{-2\pi i n x}\, dx = \int_{-1/2}^{1/2} h(x)\left(e^{2\pi i x} - 1\right) e^{-2\pi i n x}\, dx$$

$$= \int_{-1/2}^{1/2} h(x)\, e^{-2\pi i (n-1) x}\, dx - \int_{-1/2}^{1/2} h(x)\, e^{-2\pi i n x}\, dx = \widehat{h}(n-1) - \widehat{h}(n)\,.$$

Then, as $N \to +\infty$,

$$S_N g(0) = \sum_{n=-N}^{+N} \widehat{g}(n) = \sum_{n=-N}^{+N} \left(\widehat{h}(n-1) - \widehat{h}(n)\right)$$

$$= \left(\widehat{h}(-N-1) - \widehat{h}(-N)\right) + \left(\widehat{h}(-N) - \widehat{h}(-N+1)\right) + \ldots + \left(\widehat{h}(N-1) - \widehat{h}(N)\right)$$

$$= \widehat{h}(-N-1) - \widehat{h}(N) \longrightarrow 0\,.$$

Then $S_N f(0) = S_N g(0) \to 0$.   □

In particular, if $f \in C(\mathbb{T})$ is piecewise smooth we have, as $N \to +\infty$,

$$S_N f(x) \longrightarrow f(x)$$

for every $x \in \mathbb{T}$.

The above result is false when we only assume that f is continuous.

Theorem 4.19   *There exists $f \in C(\mathbb{T})$ such that*

$$\limsup_{N \to +\infty} |S_N f(0)| = +\infty\,. \tag{4.25}$$

We need the following result.

Lemma 4.20   *There exists a positive constant τ such that for every $N \in \mathbb{N}$ and $x \in \mathbb{R}$ we have*

$$\left| \sum_{j=1}^{N} \frac{\sin(2\pi j x)}{j} \right| < \tau\,. \tag{4.26}$$

Proof   Observe that for every $x \notin \mathbb{Z}$ we have

$$\sum_{n=-N}^{N} e^{2\pi i n x} = e^{-2\pi i N x} \sum_{m=0}^{2N} e^{2\pi i m x} = e^{-2\pi i N x} \sum_{m=0}^{2N} \left(e^{2\pi i x}\right)^m \tag{4.27}$$

$$= e^{-2\pi i N x} \frac{e^{2\pi i (2N+1) x} - 1}{e^{2\pi i x} - 1} = \frac{e^{2\pi i (N+\frac{1}{2}) x} - e^{-2\pi i (N+\frac{1}{2}) x}}{e^{\pi i x} - e^{-\pi i x}} = \frac{\sin\left(2\pi \left(N + \frac{1}{2}\right) x\right)}{\sin(\pi x)}\,.$$

In order to prove (4.26) it is enough to consider $x \in (0, 1/2)$. By (4.27) we have

$$\sum_{j=1}^{N} \frac{\sin(2\pi j x)}{j} = 2\pi \int_0^x \sum_{j=1}^{N} \cos(2\pi j y)\, dy = -\pi x + \pi \int_0^x \sum_{j=-N}^{N} e^{2\pi i j y}\, dy$$

$$= -\pi x + \pi \int_0^x \frac{\sin\left((2N+1)\pi y\right)}{\sin\left(\pi y\right)}\, dy$$

$$= -\pi x + \pi \int_0^x \sin\left((2N+1)\pi y\right)\left(\frac{1}{\pi y} + \left(\frac{1}{\sin\left(\pi y\right)} - \frac{1}{\pi y}\right)\right)\, dy\ .$$

Since the function $y \mapsto \left(\frac{1}{\sin(\pi y)} - \frac{1}{\pi y}\right)$ is bounded on $(0, 1/2)$, we only have to estimate

$$\int_0^x \frac{\sin\left((2N+1)\pi y\right)}{y}\, dy = \int_0^{(2N+1)\pi x} \frac{\sin u}{u}\, du\ .$$

It is enough to bound $\int_1^r \frac{\sin u}{u}\, du$ for every $r \geq 1$. Integration by parts gives

$$\left| \int_1^r \frac{\sin u}{u}\, du \right| = \left| \left[-\frac{\cos u}{u}\right]_1^r - \int_1^r \frac{\cos u}{u^2}\, du \right| \leq 2 + \int_1^{+\infty} \frac{1}{u^2}\, du = 3\ .$$

$\square$

Proof of Theorem 4.19 For positive integers M and N let

$$P_{M,N}(x) := e^{2\pi i M x} \sum_{j=1}^{N} \frac{\sin\left(2\pi j x\right)}{j}\ .$$

By (4.26) we have, for every M, N, x,

$$\left| P_{M,N}(x) \right| \leq \tau\ , \tag{4.28}$$

$$\widehat{P}_{M,N}(n) = 0 \quad \text{if} \quad n < M - N \quad \text{or} \quad n > M + N\ .$$

Let S_M denote the Fourier partial sum of degree M. Since

$$S_M P_{M,N}(x) = e^{2\pi i M x} \sum_{j=-N}^{-1} \frac{1}{2ij} e^{2\pi i j x}\ ,$$

we have

$$\left| S_M P_{M,N}(0) \right| = \frac{1}{2} \sum_{n=1}^{N} \frac{1}{n} \geq \frac{1}{2} \log(N)\ . \tag{4.29}$$

We choose two positive sequences M_j and N_j such that for every positive integer j we have

$$M_j + N_j < M_{j+1} - N_{j+1} \tag{4.30}$$

$$\log\left(N_j\right) > 2j^3 \tag{4.31}$$

Let

$$f(x) = \sum_{j=1}^{+\infty} \frac{1}{j^2} P_{M_j, N_j}(x)\ . \tag{4.32}$$

The uniform convergence of the series in (4.32) implies the continuity of f. By (4.30) the functions $n \mapsto \widehat{P_{M_j,N_j}}(n)$ and $n \mapsto \widehat{P_{M_k,N_k}}(n)$ have disjoint supports provided $j \neq k$. By (4.28), (4.29), (4.31) and (4.32) we have

$$\left| S_{M_h} f(0) \right| = \left| \sum_{j=1}^{h-1} \frac{1}{j^2} P_{M_j,N_j}(0) + \frac{1}{h^2} S_{M_h} P_{M_h,N_h}(0) \right| \geq h - c \longrightarrow +\infty$$

This proves (4.25). □

We now show some applications of Fourier series to the summation of numerical series.

Let $s : \mathbb{R} \longrightarrow \mathbb{R}$ be the *sawtooth function*

$$s(x) := \{x\} - \frac{1}{2} \tag{4.33}$$

(where $\{x\} = x - [x]$ is the fractional part of x). Note that $s(x)$ has period 1 and graph

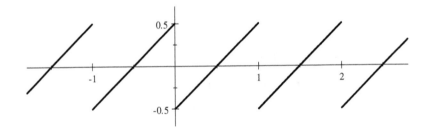

The sawtooth function is piecewise smooth and has Fourier coefficients

$$\widehat{s}(n) = \begin{cases} \int_0^1 \left(x - \frac{1}{2}\right) e^{-2\pi i n x} dx = \frac{-1}{2\pi i n} & \text{if } n \neq 0, \\ 0 & \text{if } n = 0. \end{cases}$$

Then Theorem 4.18 implies

$$\sum_{\substack{n=-\infty \\ n \neq 0}}^{+\infty} \frac{-1}{2\pi i n} e^{2\pi i n x} = -\frac{1}{\pi} \sum_{n=1}^{+\infty} \frac{\sin(2\pi n x)}{n} = \begin{cases} s(x) & \text{if } x \notin \mathbb{Z}, \\ 0 & \text{if } x \in \mathbb{Z}. \end{cases} \tag{4.34}$$

Then, for every $0 < x < 1$,

$$s(x) = x - \frac{1}{2} = -\frac{1}{\pi} \sum_{n=1}^{+\infty} \frac{\sin(2\pi n x)}{n} \, .$$

Letting $x = 1/4$ we obtain the identity

$$\frac{\pi}{4} = 1 - \frac{1}{3} + \frac{1}{5} - \frac{1}{7} + \frac{1}{9} - \cdots$$

Now observe that Parseval's identity (4.23) implies

$$\sum_{\substack{n=-\infty \\ n \neq 0}}^{+\infty} \left| \frac{-1}{2\pi i n} \right|^2 = \frac{1}{2\pi^2} \sum_{n=1}^{+\infty} \frac{1}{n^2} = \int_0^1 \left(x - \frac{1}{2} \right)^2 dx = \int_{-1/2}^{1/2} t^2 dt = \frac{1}{12} \, .$$

In this way we obtain the amazing identity

$$\sum_{n=1}^{+\infty} \frac{1}{n^2} = \frac{\pi^2}{6} \, , \tag{4.35}$$

proved by Euler in 1735 and previously known as the *Basel problem* (originally proposed by Mengoli in 1644).

4.4 Another proof of the quadratic reciprocity law

We need the following lemma on Gauss sums.

Lemma 4.21 *Let q be an odd positive number. Then*

$$S(q, 1) = \epsilon_q q^{1/2} \, ,$$

where

$$\epsilon_q = \begin{cases} 1 & \text{if } q \equiv 1 \pmod 4 \, , \\ i & \text{if } q \equiv 3 \pmod 4 \, . \end{cases} \tag{4.36}$$

Proof For every $x \in [0, 1]$ let

$$f(x) = \sum_{n=0}^{q-1} e^{2\pi i (n+x)^2/q} \, .$$

Since $e^{2\pi i(0/q)} = e^{2\pi i q^2/q}$ we have

$$f(0) = \sum_{n=0}^{q-1} e^{2\pi i n^2/q} = \sum_{n=-1}^{q-2} e^{2\pi i(n+1)^2/q} = \sum_{n=0}^{q-1} e^{2\pi i(n+1)^2/q} = f(1) \, . \tag{4.37}$$

Then the periodic continuation of $f(x)$ is continuous and piecewise smooth on $[0, 1]$. We compute the Fourier coefficients

$$\widehat{f}(k) = \int_0^1 \left(\sum_{n=0}^{q-1} e^{2\pi i(n+x)^2/q} \right) e^{-2\pi i k x} dx = \sum_{n=0}^{q-1} \int_n^{n+1} e^{2\pi i y^2/q} e^{-2\pi i k(y-n)} dy$$

$$= \sum_{n=0}^{q-1} \int_n^{n+1} e^{2\pi i\left(\frac{y^2}{q}-ky\right)} dy = \int_0^q e^{2\pi i\left(\frac{y^2}{q}-ky\right)} dy = q \int_0^1 e^{2\pi i(qu^2-kqu)} du$$

$$= qe^{-2\pi iq\frac{k^2}{4}} \int_0^1 e^{2\pi iq\left(u-\frac{k}{2}\right)^2} du = qe^{-2\pi iq\frac{k^2}{4}} \int_{-\frac{k}{2}}^{1-\frac{k}{2}} e^{2\pi iqv^2} dv \;.$$

By Definition 4.12, (4.37) and Theorem 4.18 we have

$$S(q,1) = f(0) = \lim_{N\to+\infty} \sum_{k=-N}^{+N} \widehat{f}(k) = q \lim_{N\to+\infty} \sum_{k=-N}^{+N} e^{-2\pi iq\frac{k^2}{4}} \int_{-\frac{k}{2}}^{1-\frac{k}{2}} e^{2\pi iqv^2} dv$$

$$= q \lim_{N\to+\infty} \left(\sum_{\substack{k=-N \\ k \text{ even}}}^{+N} \int_{-\frac{k}{2}}^{1-\frac{k}{2}} e^{2\pi iqv^2} dv + e^{-2\pi iq/4} \sum_{\substack{k=-N \\ k \text{ odd}}}^{+N} \int_{-\frac{k}{2}}^{1-\frac{k}{2}} e^{2\pi iqv^2} dv \right).$$

Since

$$\int_R^{+\infty} e^{2\pi iv^2} dv = \int_{R^2}^{+\infty} \frac{e^{2\pi it}}{2t^{1/2}} dt = \left[\frac{e^{2\pi it}}{4\pi it^{1/2}}\right]_{R^2}^{+\infty} + \int_{R^2}^{+\infty} \frac{e^{2\pi it}}{8\pi it^{3/2}} dt \longrightarrow 0$$

as $|R| \to +\infty$, we obtain

$$\lim_{N\to+\infty} \sum_{\substack{k=-N \\ k \text{ even}}}^{+N} \int_{-\frac{k}{2}}^{1-\frac{k}{2}} e^{2\pi iqv^2} dv = \lim_{N\to+\infty} \sum_{\substack{k=-N \\ k \text{ odd}}}^{+N} \int_{-\frac{k}{2}}^{1-\frac{k}{2}} e^{2\pi iqv^2} dv = \int_{-\infty}^{+\infty} e^{2\pi iqv^2} dv \;,$$

where

$$\int_{-\infty}^{+\infty} e^{2\pi iqv^2} dv := \lim_{R\to+\infty} \int_{-R}^{R} e^{2\pi iqv^2} dv \;.$$

Then

$$S(q,1) = q\left(1 + e^{-\pi iq/2}\right) \int_{-\infty}^{+\infty} e^{2\pi iqv^2} dv = \sqrt{q}\left(1 + i^{-q}\right) \int_{-\infty}^{+\infty} e^{2\pi is^2} ds \;.$$

If we choose $q = 1$ and observe that $S(1,1) = 1$, then (without use of complex integration[2]) we obtain

$$\int_{-\infty}^{+\infty} e^{2\pi iv^2} dv = \frac{1}{1-i} \;. \tag{4.38}$$

Then

$$S(q,1) = \frac{\sqrt{q}\left(1 + i^{-q}\right)}{1-i} \;.$$

[2] For a complex analytic proof of (4.38) see e.g. [156, p. 181]. For a real analytic proof see [140, 2.7.3].

Since

$$\frac{1+i^{-q}}{1-i} = \begin{cases} 1 & \text{if } q \equiv 1 \pmod 4, \\ i & \text{if } q \equiv 3 \pmod 4, \end{cases}$$

we complete the proof. □

Second proof of Theorem 4.11 Let p, q be two distinct prime numbers. Then Lemmas 4.13 and 4.14 imply

$$S(pq, 1) = S(p, q)S(q, p) = \left(\frac{q}{p}\right)_L S(p, 1) \left(\frac{p}{q}\right)_L S(q, 1).$$

By Lemma 4.21 and the notation in (4.36) we deduce that

$$\left(\frac{p}{q}\right)_L \left(\frac{q}{p}\right)_L = \frac{S(pq, 1)}{S(p, 1)S(q, 1)} = \frac{\epsilon_{pq}}{\epsilon_p \epsilon_q}.$$

Finally observe that if $p \equiv q \equiv 3 \pmod 4$, then $pq \equiv 1 \pmod 4$ and $\frac{\epsilon_{pq}}{\epsilon_p \epsilon_q} = -1$, while the other values of p and q give $\frac{\epsilon_{pq}}{\epsilon_p \epsilon_q} = 1$. □

Exercises

1) Do the following congruences have solutions?

$$x^2 \equiv 219 \pmod{383}, \quad x^2 \equiv 650 \pmod{1109}, \quad x^2 \equiv 611 \pmod{1009}.$$

2) Does the congruence $x^2 \equiv 196 \pmod{1357}$ have a solution? (Observe that 1357 is not a prime number.)

3) Use the identity

$$\left(\frac{-1}{p}\right)_L = \begin{cases} 1 & \text{if } p \equiv 1 \pmod 4, \\ -1 & \text{if } p \equiv 3 \pmod 4 \end{cases}$$

to deduce another proof of Theorem 3.10.

4) Let p be an odd prime number. Prove the following identities:

$$\sum_{a=1}^{p-1} \left(\frac{a}{p}\right)_L = 0, \quad \sum_{a=1}^{p-2} \left(\frac{a}{p}\right)_L = (-1)^{(p-1)/2}.$$

5) Let u_n and v_n be two sequences in a Hilbert space $\mathcal{H}$. Assume that $u_n \to u$ and $v_n \to v$. Prove that $\langle u_n, v_n \rangle \to \langle u, v \rangle$.

6) Let $f(x)$ have period 1 on $\mathbb{R}$ and satisfy $g(x) = x^2$ when $x \in [0, 1)$. Write the Fourier series of f and find N such that $\|S_N f - f\|_{L^2(\mathbb{T})} \le \frac{1}{10}$.

7) Let $f \in L^2(\mathbb{T})$ and $Q_N(x) = \sum_{k=-N}^N c_k e^{2\pi i k x}$. Prove that

$$\|f - S_N f\|_{L^2(\mathbb{T})} \le \|f - Q_N\|_{L^2(\mathbb{T})}.$$

8) Let $f \in L^1(\mathbb{T})$ and $g \in L^\infty(\mathbb{T})$. Prove that

$$\lim_{n \to +\infty} \int_0^1 f(t)\, g(nt)\, dt = \widehat{f}(0)\, \widehat{g}(0) \ .$$

9) Let $f \in C(\mathbb{T})$ be piecewise smooth on (say) $[0, 1]$. Prove that the Fourier series of f converges absolutely (that is, $\sum_{n \in \mathbb{Z}} |\widehat{f}(n)| < +\infty$) and deduce that $S_N f(x)$ converges uniformly to $f(x)$.

10) Let f be piecewise smooth on $\mathbb{T}$. Prove that the Fourier series of f converges uniformly to $f(x)$ in every closed interval where f is continuous.

11) Compute $\sum_{n=1}^{+\infty} \frac{1}{n^4}$.

5

Sums of squares

In this chapter we first determine the integers which can be written as sums of two squares, then we shall prove a celebrated result of Lagrange which says that every positive integer can be written as a sum of four squares. In geometric terms, Lagrange's theorem says that for every positive integer n, we have

$$\left\{t \in \mathbb{R}^4 : |t| = \sqrt{n}\right\} \cap \mathbb{Z}^4 \neq \varnothing .$$

That is, every 4-dimensional sphere with centre 0 and radius $\sqrt{n}$ contains at least one integer point.

In the first section we introduce two elegant results with a wide range of applications: Minkowski's theorem and Dirichlet's theorem.

5.1 The theorems of Minkowski and Dirichlet

We need to define lattices in $\mathbb{R}^d$ (the most familiar example is $\mathbb{Z}^d$).

Definition 5.1 Let $\{p_1, \ldots, p_d\}$ be a basis of $\mathbb{R}^d$. By a *lattice* we mean the additive group

$$L = \left\{\sum_{j=1}^{d} m_j p_j\right\}_{m_1, \ldots, m_d \in \mathbb{Z}}$$

generated by the above basis. The closed set

$$K_L := \left\{\sum_{j=1}^{d} x_j p_j : 0 \leq x_j \leq 1, \ j = 1, \ldots, d\right\} \tag{5.1}$$

and its translates in L are called *fundamental parallelepipeds* of L, and we write $A(L)$ for their common volume.

Observe that the union

$$\bigcup_{p \in L} (K_L + p) = \mathbb{R}^d$$

is disjoint up to sets of measure zero.

The following result has been proved by Minkowski in 1889. It concerns *convex bodies* in $\mathbb{R}^d$, that is, convex sets with positive and finite measure.

Theorem 5.2 (Minkowski) *Let L be a lattice in $\mathbb{R}^d$.*

(i) *Let $D \subset \mathbb{R}^d$ be a convex body, symmetric around the origin (i.e., $x \in D$ implies $-x \in D$), with volume $|D| > 2^d A(L)$. Then D contains points of L other than 0.*

(ii) *Let $D \subset \mathbb{R}^d$ be a closed convex body, symmetric around the origin, with volume $|D| \geq 2^d A(L)$. Then D contains points of L other than 0.*

Proof (i) We first claim the existence of two different points $x, y \in D$ such that $\frac{1}{2}x - \frac{1}{2}y \in L$. Indeed, for every $p \in L$ we have

$$D_p := \left(\left(\frac{1}{2}D \right) \cap (K_L + p) \right) - p \subseteq K_L ,$$

and therefore $\cup_{p \in L} D_p \subseteq K_L$. Then

$$\sum_{p \in L} |D_p| = \left| \frac{1}{2}D \right| = 2^{-d} |D| > A(L) = |K_L| \geq \left| \bigcup_{p \in L} D_p \right| .$$

Hence the union $\cup_{p \in L} D_p$ is not disjoint and there are $p, q \in L$ and $x, y \in D$ ($x \neq y$) such that $\frac{1}{2}x - p = \frac{1}{2}y - q$. Observe that $p \neq q$. By symmetry and convexity we have

$$0 \neq p - q = \frac{1}{2}x - \frac{1}{2}y = \frac{x + (-y)}{2} \in D .$$

(ii) Now D is a closed set. We may assume $|D| = 2^d A(L)$. Again we claim that the union $\cup_{p \in L} D_p$ is not disjoint. Assume the contrary. If we have $\left| \cup_{p \in L} D_p \right| < A(L)$, we proceed as before. Then we assume

$$\left| \bigcup_{p \in L} D_p \right| = 2^{-d} |D| = A(L) .$$

We observe that D is a bounded set. Assume the contrary. Since $|D| > 0$ there exist $n + 1$ points $x_1, \ldots, x_{n+1}$ in D which are not coplanar. If D is unbounded there exists $y \in D$ far away from a ball containing $x_1, \ldots, x_{n+1}$. Then the convex hull of $x_1, \ldots, x_{n+1}, y$ (that is, the intersection of all convex sets containing $x_1, \ldots, x_{n+1}, y$) is contained in D and has large volume. Hence D is bounded and the set $\cup_{p\in L}D_p$ is the disjoint union of a finite number of closed sets such that

$$\bigcup_{p\in L}D_p \subseteq K_L, \qquad \left|\bigcup_{p\in L}D_p\right| = |K_L|.$$

This is impossible because the sets D_p are closed pairwise disjoint sets, therefore they must have positive distance from one another. □

Minkowski's theorem has several interesting applications. Among them is *Pick's theorem* (see [126]), which says that a convex polygon P (with interior P^o and boundary ∂P) having vertices in $\mathbb{Z}^2$ has area

$$|P| = \text{card}\left(P^o \cap \mathbb{Z}^2\right) + \frac{1}{2}\,\text{card}\left(\partial P \cap \mathbb{Z}^2\right) - 1.$$

We now introduce Dirichlet's theorem.

We know that $\mathbb{Q}$ is dense in $\mathbb{R}$, and it is easy to show that for every $\alpha \in \mathbb{R}$ and $q \in \mathbb{N}$ there exists $p \in \mathbb{Z}$ such that $\left|\alpha - \frac{p}{q}\right| \le \frac{1}{2q}$. Indeed, this means that $|\alpha q - p| \le 1/2$ and this is obvious, since for every number αq there exists an integer which is at most a distance $1/2$ away. The following result, proved by Dirichlet in 1840, is far more interesting.

Theorem 5.3 (Dirichlet) *For all $\alpha \in \mathbb{R}$ and $N \in \mathbb{N}$ there exist $p, q \in \mathbb{Z}$ such that $1 \le q \le N$ and $|\alpha q - p| \le 1/(N+1)$.*

Proof The $N + 2$ numbers

$$0, \{\alpha\}, \{2\alpha\}, \ldots, \{N\alpha\}, 1$$

belong to the interval $[0, 1]$ and, as usual, we write $[\beta]$ for the integral part of a real number β and $\{\beta\} = \beta - [\beta]$ for the fractional part. Then[1] at least two of them (termed r and s, respectively) satisfy $|r - s| \le 1/(N+1)$. If $r = 0$ and $s = \{k\alpha\}$, then

$$0 \le k\alpha - [k\alpha] = \{k\alpha\} \le \frac{1}{N+1}$$

and we choose $q = k$ and $p = [k\alpha]$. The case $r = \{k\alpha\}$ and $s = 1$ is similar. Then we may assume that $r = \{k\alpha\}$ and $s = \{\ell\alpha\}$, with $0 < k < \ell$. We have

$$|(\ell - k)\alpha - [\ell\alpha] + [k\alpha]| = |\{\ell\alpha\} - \{k\alpha\}| \le \frac{1}{N+1}.$$

Then we choose $q = \ell - k$ and $p = [\ell\alpha] - [k\alpha]$. This completes the proof. □

---

[1] This easy argument has been popularized by Dirichlet as the *pigeonhole principle*. It says that if $n + 1$ pigeons have to be put in n holes, then at least one hole will contain more than one pigeon.

Corollary 5.4 *For every irrational number α there exist infinitely many rational numbers p/n satisfying*

$$\left| \alpha - \frac{p}{n} \right| < \frac{1}{n^2} . \tag{5.2}$$

Proof Given a positive integer N_1, the previous theorem implies the existence of at least one rational number p_1/n_1 such that

$$\left| \alpha - \frac{p_1}{n_1} \right| < \frac{1}{n_1 N_1} \le \frac{1}{n_1^2} .$$

Since α is irrational, there exists a positive integer N_2 such that $|\alpha n_1 - p_1| > \frac{1}{N_2}$. We now repeat the previous argument with N_2 in place of N_1. In this way we find $\frac{p_2}{n_2} \ne \frac{p_1}{n_1}$ such that $\left| \alpha - \frac{p_2}{n_2} \right| \le \frac{1}{n_2^2}$. And so we go on. $\square$

Remark 5.5 Corollary 5.4 is no longer true if $\alpha \in \mathbb{Q}$. Assume the contrary, then there exist integers $a \ge 0$ and $b > 0$ such that for infinitely many pairwise different rational numbers p_j/q_j we have

$$\left| \frac{a}{b} - \frac{p_j}{q_j} \right| < \frac{1}{q_j^2} .$$

We may assume $q_j > 0$ and $p_j/q_j \ne a/b$. Observe that the sequence q_j is unbounded. Since

$$0 \ne \left| \frac{a}{b} - \frac{p_j}{q_j} \right| = \frac{|aq_j - bp_j|}{bq_j} \ge \frac{1}{bq_j} ,$$

we obtain a contradiction.

The following result shows that the exponent 2 in Corollary 5.4 cannot be improved.

Theorem 5.6 *Let α be an irrational algebraic number of degree 2 (for example, $\alpha = \sqrt{2}$). Then there exists a positive constant H such that for every rational number p/n we have*

$$\left| \alpha - \frac{p}{n} \right| \ge \frac{H}{n^2} .$$

Proof We may assume n large. There exists a second-degree polynomial $Q(x) = Mx^2 + Nx + R$ with integral coefficients, such that $Q(\alpha) = 0$. By the mean value theorem there exists θ between α and p/n such that

$$\frac{Q(\alpha) - Q\left(\frac{p}{n}\right)}{\alpha - \frac{p}{n}} = Q'(\theta) . \tag{5.3}$$

Since α is irrational, we have $Q'(\alpha) = 2M\alpha + N \neq 0$. Hence there is $\delta > 0$ such that $|Q'(x)| < 2|Q'(\alpha)|$ for $|x - \alpha| < \delta$. We may assume $|\alpha - p/n| < 1/n^2$, otherwise there is nothing to prove. Since n is large we have $|\alpha - p/n| < \delta$ and $|\alpha - \theta| < \delta$. Then (5.3) implies

$$\left|\alpha - \frac{p}{n}\right| = \left|\frac{Q\left(\frac{p}{n}\right)}{Q'(\theta)}\right| > \left|\frac{Q\left(\frac{p}{n}\right)}{2Q'(\alpha)}\right| = \frac{1}{2|Q'(\alpha)|}\left|M\left(\frac{p}{n}\right)^2 + N\left(\frac{p}{n}\right) + R\right|$$

$$= \frac{1}{2|Q'(\alpha)|}\frac{\left|Mp^2 + Npn + Rn^2\right|}{n^2} \geq \frac{1}{2|Q'(\alpha)|n^2} .$$

Indeed, $Q\left(\frac{p}{n}\right) \neq 0$ and then $Mp^2 + Npn + Rn^2$ is a non-zero integer. □

Hurwitz proved in 1891 (see [76, Ch. 4] or [90, Ch. XI]) that (5.2) can be improved up to the inequality

$$\left|\alpha - \frac{p}{n}\right| < \frac{1}{\sqrt{5}n^2} . \tag{5.4}$$

The following result shows that (5.4) cannot be improved further (see e.g. [151]).

Theorem 5.7 *Let $L > \sqrt{5}$. Then only a finite number of rational numbers p/q satisfy the inequality*

$$\left|\frac{p}{q} - \frac{\sqrt{5}-1}{2}\right| < \frac{1}{Lq^2} . \tag{5.5}$$

Proof Assume that p/q satisfies (5.5) for some $L > \sqrt{5}$. Let

$$f(x) = x^2 + x - 1 = \left(x - \frac{\sqrt{5}-1}{2}\right)\left(x + \frac{\sqrt{5}+1}{2}\right) .$$

Since $f(p/q) \neq 0$, (5.5) implies

$$\frac{1}{q^2} \leq \left|\frac{p^2 + pq - q^2}{q^2}\right| = \left|\left(\frac{p}{q}\right)^2 + \left(\frac{p}{q}\right) - 1\right| = \left|f\left(\frac{p}{q}\right)\right|$$

$$= \left|\frac{p}{q} - \frac{\sqrt{5}-1}{2}\right|\left|\frac{p}{q} + \frac{\sqrt{5}+1}{2}\right| \leq \frac{1}{Lq^2}\left|\frac{p}{q} - \frac{\sqrt{5}-1}{2} + \sqrt{5}\right|$$

$$\leq \frac{1}{Lq^2}\left(\frac{1}{Lq^2} + \sqrt{5}\right) .$$

Hence

$$q^2 \leq \frac{1}{L\left(L - \sqrt{5}\right)} .$$

Then only finitely many rational numbers p/q satisfy (5.5). □

5.2 Sums of two squares

Let us go back to the arithmetic function $r(n)$ introduced in (2.14). The following result is due to Girard and Fermat. We will give two proofs, one based on Dirichlet's theorem and another based on Minkowski's theorem.

Theorem 5.8 (Girard and Fermat) *Let $p \equiv 1 \pmod 4$ be a prime number. Then $r(p) \geq 1$.*

First proof By Remark 4.7 we know that the congruence

$$z^2 + 1 \equiv 0 \pmod p \tag{5.6}$$

has solutions. By Theorem 5.3 there exists w/s such that $1 \leq s \leq \left[\sqrt{p}\right]$ and

$$\left| \frac{z}{p} - \frac{w}{s} \right| \leq \frac{1}{\left(\left[\sqrt{p}\right]+1\right)s} \ .$$

Writing $t = zs - wp$ we have

$$|t| \leq \frac{p}{\left[\sqrt{p}\right]+1} < \sqrt{p} \ . \tag{5.7}$$

Then

$$s^2 + t^2 = s^2 + (zs - wp)^2 = s^2\left(1 + z^2\right) + p\left(w^2 p - 2zsw\right) \ .$$

By (5.6) p divides the above RHS, so that $p \mid \left(s^2 + t^2\right)$. By (5.7) and the inequality $1 \leq s \leq \sqrt{p}$ we obtain $0 < s^2 + t^2 < 2p$. Then $p = s^2 + t^2$. □

Second proof Let z be as in (5.6). We consider the lattice

$$L = \left\{(x, y) \in \mathbb{Z}^2 : y \equiv zx \pmod p\right\}$$

generated by $(1, z)$ and $(0, p)$. We have $A(L) = p$. The open disc

$$B^o := \left\{(x, y) \in \mathbb{R}^2 : x^2 + y^2 < 2p\right\}$$

centred at 0 and with radius $\sqrt{2p}$ has area $2\pi p > 2^2 p$. Then Minkowski's theorem implies that B^o contains a point of L other than the origin. In other words, there exists $(x, y) \in L$ such that $0 < x^2 + y^2 < 2p$. As in the first proof, note that (5.6) implies

$$x^2 + y^2 = x^2\left(1 + z^2\right) \equiv 0 \pmod p \ .$$

Hence $x^2 + y^2 = p$. □

The following theorem was proved by Jacobi in 1834 and provides a simple way to decide whether a given positive integer can be written as a sum of two squares.

Theorem 5.9 (Jacobi) *Let an integer $n > 1$ have canonical decomposition*

$$n = 2^r p_1^{r_1} \cdots p_k^{r_k} q_1^{s_1} \cdots q_\ell^{s_\ell} , \tag{5.8}$$

where $p_1 \equiv \ldots \equiv p_k \equiv 1 \pmod{4}$ and $q_1 \equiv \ldots \equiv q_\ell \equiv 3 \pmod{4}$. Then n can be written as a sum of two squares if and only if every s_m is even.

Proof Let $n = x^2 + y^2$ and let $d = (x, y)$ (the greatest common divisor of x and y). Assume (say) that s_1 is odd and let

$$\widetilde{n} = \frac{n}{d^2} = \left(\frac{x}{d}\right)^2 + \left(\frac{y}{d}\right)^2 := x_0^2 + y_0^2 .$$

Then q_1 appears with an odd exponent in the canonical decomposition of $\widetilde{n}$. Hence $q_1 \mid (x_0^2 + y_0^2)$. Then if $q_1 \mid x_0^2$ we also have $q_1 \mid y_0^2$. Since $(x_0, y_0) = 1$, we have $(q_1, x_0^2) = (q_1, y_0^2) = 1$. Then Theorem 3.5 implies the existence of an integer z such that $y_0 z \equiv x_0 \pmod{q_1}$. Hence

$$0 \equiv \widetilde{n} \equiv x_0^2 + y_0^2 \equiv y_0^2 \left(1 + z^2\right) \pmod{q_1} .$$

Then $z^2 \equiv -1 \pmod{q_1}$, against Remark 4.7. Conversely, let us write

$$n = 2^r p_1^{r_1} \cdots p_k^{r_k} q_1^{s_1} \cdots q_\ell^{s_\ell} , \tag{5.9}$$

where

$$p_1 \equiv \ldots \equiv p_k \equiv 1 \pmod{4} , \qquad q_1 \equiv \ldots \equiv q_\ell \equiv 3 \pmod{4}$$

and every s_h is even. Since

$$\left(a^2 + b^2\right)\left(A^2 + B^2\right) = (aA - bB)^2 + (aB + bA)^2 , \tag{5.10}$$

it is sufficient to prove that every prime power factor in the product (5.9) is a sum of two squares. Indeed $2 = 1 + 1$, each p_j is a sum of two squares by Theorem 5.8 and every $q_m^{s_m} = q_m^{s_m} + 0$ is trivially a sum of two squares since s_m is even. This completes the proof. $\square$

Example 5.10 Let us consider the number $12103 = 19 \cdot 7^2 \cdot 13$. The prime number $19 \equiv 3 \pmod{4}$ has an odd power, so 12103 cannot be written as a sum of two squares.

Example 5.11 Let us consider the number $196625 = 11^2 \cdot 5^3 \cdot 13$, where 11 is the only prime factor $\equiv 3 \pmod{4}$. Its power is even, so 196625 can be written as a sum of two squares. Let us find a way of doing this. Observe that

$$11^2 = 11^2 + 0^2 , \qquad 5^3 = 10^2 + 5^2 , \qquad 13 = 3^2 + 2^2$$

and then, by the argument in (5.10),

$$196625 = \left(11^2 + 0^2\right)\left(10^2 + 5^2\right)\left(3^2 + 2^2\right)$$
$$= \left((11 \cdot 10 + 0 \cdot 5)^2 + (11 \cdot 5 - 0 \cdot 10)^2\right)\left(3^2 + 2^2\right) = \left(110^2 + 55^2\right)\left(3^2 + 2^2\right)$$
$$= (110 \cdot 3 + 55 \cdot 2)^2 + (110 \cdot 2 - 55 \cdot 3)^2 = 440^2 + 55^2 \ .$$

The identity (5.10) is clearer in terms of complex numbers. Indeed, let us write

$$z = a + ib \ , \quad Z = A + iB \ .$$

Then

$$|z|^2 = a^2 + b^2 \ , \quad |Z|^2 = A^2 + B^2 \ , \quad zZ = (aA - bB) + i(aB + bA)$$

and (5.10) is nothing but

$$|z|^2 |Z|^2 = |zZ|^2 \ .$$

The above simple remark introduces the following section.

5.3 Gaussian integers

In this section we shall extend Theorem 5.9 by establishing an explicit identity for $r(n)$. This will be done by replacing $\mathbb{Z}$ with the larger ring

$$\mathbb{Z}[i] = \{z \in \mathbb{C} : z = x + iy : x, y \in \mathbb{Z}\}$$

of *Gaussian integers*. $\mathbb{Z}[i]$ can be identified with $\mathbb{Z}^2$, addition and multiplication come from $\mathbb{C}$, as well as conjugation and absolute value. Actually $\mathbb{Z}[i]$ is an integral domain, that is, a commutative ring with 1 that has no zero divisors (that is, $ab = 0$ implies $a = 0$ or $b = 0$). It will be useful to define the function $N(z) = |z|^2$, which takes integral values. Observe that

$$N(wz) = N(w)N(z) \ .$$

$\mathbb{Z}[i]$ is the simplest extension of $\mathbb{Z}$ and we shall see that several results from Section 1.1 still hold true for $\mathbb{Z}[i]$.

Proposition 5.12 *Let $w \neq 0$ and z be Gaussian integers. Then there exist $q, r \in \mathbb{Z}[i]$ such that $z = qw + r$ and $N(r) < N(w)$.*

Proof Since $w \neq 0$, the set $\{qw\}_{q \in \mathbb{Z}[i]}$ is a lattice generated by w and iw, with squares of side length $|w|$ as fundamental parallelograms. Then for every $z \in \mathbb{Z}[i]$ there exists $q \in \mathbb{Z}[i]$ such that $|z - qw| \leq |w| / \sqrt{2}$. See the figure below.   □

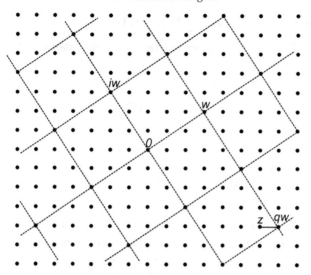

If $r = 0$, we say that w is a divisor of z and write $w \mid z$.

The four Gaussian integers ± 1, $\pm i$ are the only elements with absolute value 1 and are the only invertible elements in $\mathbb{Z}[i]$. They are called *units*. We say that two elements $a, b \in \mathbb{Z}[i]$ are *associated* if there is a unit u such that $a = ub$. Observe that the elements associated with $x + iy$ are

$$(x + iy) , \quad (-x - iy) , \quad (-y + ix) , \quad (y - ix) ,$$

while in general $x - iy$ is not associated with $x + iy$. An element $0 \neq \pi \in \mathbb{Z}[i]$ which is not a unit is called[2] a *Gaussian prime* if its only divisors are units or elements associated with π, that is, if $\pi = ab$ implies $N(a) = 1$ or $N(b) = 1$.

If a positive integer is a Gaussian prime, then it is also a prime number. The converse is not true, as $2 = (1 + i)(1 - i)$ and $5 = (1 + 2i)(1 - 2i)$ are not Gaussian primes. More generally, by Theorem 5.8, every prime number $p \equiv 1 \pmod 4$ can be written as a sum of two squares:

$$p = x^2 + y^2 = (x + iy)(x - iy) ,$$

with non-zero x and y, so that p is not a Gaussian prime.

Observe that if $N(\pi)$ is a prime number, then π is a Gaussian prime. The converse is false. Indeed, $N(3) = 9$ is not a prime number, but 3 is a Gaussian prime. Indeed, assume the contrary and write $3 = ab$, with $1 < N(a) < N(3) = 9$. Then $N(a) = 3$. If $a = a_1 + ia_2$, then $3 = a_1^2 + a_2^2$, which is impossible.

---

[2] We shall not confuse the term *prime* (or *prime number*), used only for elements in $\mathbb{Z}$, with *Gaussian prime*.

The following general result will be proved later in this section.

Theorem 5.13 *The Gaussian primes are (up to conjugation and products by units)*

(i) *the prime numbers* $\equiv 3$ (mod 4),
(ii) *the Gaussian integers* π *such that* $N(\pi) = 2$ *or* $N(\pi)$ *is a prime number* $\equiv 1$ (mod 4).

First we have to show that the Gaussian integers satisfy a fundamental theorem of arithmetic (see [3, 79, 101, 120]). We need a few steps.

We start with the Euclidean algorithm for Gaussian integers. Let $w, z \in \mathbb{Z}[i]$, then Proposition 5.12 allows us to write

$$z = q_1 w + r_1 \qquad \text{with } 0 < N(r_1) < N(w)$$
$$w = q_2 r_1 + r_2 \qquad \text{with } 0 < N(r_2) < N(r_1)$$
$$r_1 = q_3 r_2 + r_3 \qquad \text{with } 0 < N(r_3) < N(r_2)$$
$$r_2 = q_4 r_3 + r_4 \qquad \text{with } 0 < N(r_4) < N(r_3)$$
$$\vdots$$
$$r_{n-3} = q_{n-1} r_{n-2} + r_{n-1} \qquad \text{with } 0 < N(r_{n-1}) < N(r_{n-2})$$
$$r_{n-2} = q_n r_{n-1}.$$

Following the lines in reverse order we see that $r_{n-1} \mid z$ and $r_{n-1} \mid w$. In a similar way we see that every divisor of z and w is also a divisor of r_{n-1}. We have therefore proved the following result (see Remark 1.5).

Theorem 5.14 *Let z and w be Gaussian integers which are not both zero. Then there exists $d \in \mathbb{Z}[i]$ such that*

(i) $d \mid z$ *and* $d \mid w$,
(ii) *if* $t \mid z$ *and* $t \mid w$, *then* $t \mid d$,
(iii) *there exist two Gaussian integers a and b such that $d = az + bw$.*

We call d a *greatest common divisor (Gaussian gcd)* of z and w. The Gaussian gcd is unique up to units. We write $d := (z, w)$.

Theorem 5.15 *Let $c \in \mathbb{Z}[i]$ satisfy $N(c) > 1$. Then c can be written as a product of Gaussian primes.*

Proof We use induction on the values of $N(c)$. If $N(c) = 2$, then (up to associated Gaussian integers) $c = 1 + i$, hence c is a Gaussian prime. Let $N(c) > 2$. If c is a Gaussian prime, the proof is complete, otherwise we have $c = ab$, with $1 < N(a) < N(c)$ and $1 < N(b) < N(c)$. Then the induction assumption implies $a = \pi_1 \cdots \pi_\ell$ and $b = \widetilde{\pi}_1 \cdots \widetilde{\pi}_m$, where $\pi_1, \ldots, \pi_\ell$ and $\widetilde{\pi}_1, \ldots, \widetilde{\pi}_m$ are Gaussian primes. Hence c is a product of Gaussian primes. $\square$

Proposition 5.16 *If $s \mid zw$ and $(s, z) = 1$, then $s \mid w$. Moreover, if $(s, z) = 1$ and $(s, z') = 1$, then $(s, zz') = 1$.*

Proof We have $zw = rs$ and $1 = as + bz$ for suitable Gaussian integers r, a, b. Then

$$w = asw + bzw = asw + brs,$$

so that $s \mid w$. As for the second part, we have $1 = as + bz$ and $1 = \alpha s + \beta z'$, hence $1 = s(a\alpha s + bz\alpha + a\beta z') + zz'b\beta$. Then $(s, zz') = 1$. □

Proposition 5.17 *Let $\pi, \pi_1, \ldots, \pi_\ell$ be Gaussian primes such that $\pi \mid \pi_1 \cdots \pi_\ell$. Then π is associated with one of the Gaussian primes $\pi_1, \ldots, \pi_\ell$.*

Proof We may assume that π is not associated with any of $\pi_1, \ldots, \pi_{\ell-1}$. Then the previous proposition implies $\pi \mid \pi_\ell$. Since π and π_ℓ are Gaussian primes we have $\pi_\ell = u\pi$, where u is a unit. □

Theorem 5.18 *Every Gaussian integer z (with $N(z) > 1$) can be written as a product of Gaussian primes. The canonical decomposition is unique up to order of factors and multiplication by units.*

Proof We only have to prove the uniqueness. Assume that

$$z = \pi_1 \ldots \pi_\ell = \widetilde{\pi}_1 \ldots \widetilde{\pi}_m.$$

Then $\pi_1 \mid \widetilde{\pi}_1 \cdots \widetilde{\pi}_m$ and the previous proposition implies that π_1 is associated with (say) $\widetilde{\pi}_1$. Hence

$$\pi_2 \ldots \pi_\ell = u \widetilde{\pi}_2 \ldots \widetilde{\pi}_m,$$

where u is an invertible element. Continuing in this way we see that $\ell = m$ and the Gaussian primes π_j and $\widetilde{\pi}_k$ are associated in pairs. □

Proof of Theorem 5.13 We prove that a prime number $p \equiv 3 \pmod 4$ is a Gaussian prime. Assume the contrary. Then $p = ab$ with $N(a) = N(b) = p$. If we write $a = x + iy$ we obtain $x^2 + y^2 = p$, contradicting Theorem 5.9. We have already seen that if $N(\pi)$ is a prime number, then π is a Gaussian prime. To complete the proof we have to show that if π is a Gaussian prime, then it satisfies one of the two conditions in the statement of the theorem. Indeed, $N(\pi) = \pi\overline{\pi}$, where the complex conjugate $\overline{\pi}$ is also a Gaussian prime. Hence the identity $N(\pi) = \pi\overline{\pi}$ represents (up to the order and up to products of units) the unique canonical decomposition of the Gaussian integer $N(\pi)$. If the integer $N(\pi)$ is a prime number and $N(\pi) = 2$ or $N(\pi) \equiv 1 \pmod 4$, then (ii) is satisfied. The case $N(\pi)$ is prime and $N(\pi) \equiv 3 \pmod 4$ is not possible, for $\pi = x + iy$ implies $N(\pi) = x^2 + y^2$, contradicting Theorem 5.9. Finally, if

$N(\pi)$ is not a prime number, we write $N(\pi) = tv$ with $t, v \in \mathbb{N}$ and not equal to 1. Note that $\pi\bar{\pi} = tv$, so it follows from uniqueness of factorization that $u\pi$ equals, say, t for a suitable unit u. As a Gaussian prime, $u\pi = t$ is also a prime number. If $u\pi = 2$ or $u\pi \equiv 1 \pmod 4$, then $u\pi$ is not a Gaussian prime. Hence $u\pi \equiv 3 \pmod 4$. □

Now we go back to the study of the arithmetic function $r(n)$.

Lemma 5.19 *Let $p \equiv 1 \pmod 4$ be a prime number. Then p is the product of two non-associated Gaussian primes and $r(p) = 8$.*

Proof We already know that p is not a Gaussian prime. We can write

$$p = x^2 + y^2 = (x + iy)(x - iy) = \pi\bar{\pi} = \bar{\pi}\pi ,$$

with $x \neq y$ and $\pi, \bar{\pi}$ Gaussian primes. By Theorem 5.18 we know that p can be written only as $p = u(x + iy)\bar{u}(x - iy)$ or $p = u(y + ix)\bar{u}(y - ix)$, where u is an arbitrary unit. Since $x \neq 0$, $y \neq 0$ and $x \neq y$, then p can be written as a sum of two squares in precisely 8 ways, that is, $r(p) = 8$. In other words we have

$$p = (\pm x)^2 + (\pm y)^2 = (\pm y)^2 + (\pm x)^2 .$$

□

We now establish an explicit identity for $r(n)$, proved by Jacobi in 1834.

Theorem 5.20 (Jacobi) *Let $n \in \mathbb{N}$ have canonical decomposition*

$$n = 2^r p_1^{r_1} \cdots p_k^{r_k} q_1^{s_1} \cdots q_\ell^{s_\ell} , \tag{5.11}$$

where $p_1 \equiv \ldots \equiv p_k \equiv 1 \pmod 4$ and $q_1 \equiv \ldots \equiv q_\ell \equiv 3 \pmod 4$. Then

$$r(n) = \begin{cases} 4 \prod_{j=1}^k (r_j + 1) & \text{if every } s_m \text{ is even,} \\ 0 & \text{otherwise.} \end{cases}$$

Proof We look at (5.11) and factor n in $\mathbb{Z}[i]$:

$$n = (1 + i)^r (1 - i)^r \prod_{j=1}^k (x_j + iy_j)^{r_j} (x_j - iy_j)^{r_j} \prod_{m=1}^\ell q_j^{s_m} ,$$

where $(x_j + iy_j)(x_j - iy_j) = x_j^2 + y_j^2 = p_j$ and Theorem 5.13 implies that the Gaussian integers $1 \pm i$, $x_j \pm iy_j$, q_j are all Gaussian primes. Observe that n is a sum of two squares if and only if there exists $z \in \mathbb{Z}[i]$ such that $n = z\bar{z}$. Hence $r(n)$ counts the number of these factors z of n in $\mathbb{Z}[i]$. We can have

$$z = (1 + i)^{r'} (1 - i)^{r - r'} \prod_{j=1}^k (x_j + iy_j)^{r'_j} (x_j - iy_j)^{r_j - r'_j} \prod_{m=1}^\ell q_j^{s_m/2} ,$$

where $0 \leq r' \leq r, 0 \leq r'_j \leq r_j$ for every $j = 1, \ldots, k$. Since $\frac{1+i}{1-i} = i$, we have to count the expressions

$$u \prod_{j=1}^{k} (x_j + iy_j)^{r'_j} (x_j - iy_j)^{r_j - r'_j} ,$$

where u is a unit. Observe that we have four choices for u and $r_j + 1$ choices for each r'_j. This completes the proof. $\qquad \square$

We now show that $r(n)$ and the Dirichlet function $d(n)$ (see (2.8)) are related by the arithmetic function

$$\delta(n) := d_1(n) - d_3(n) , \qquad (5.12)$$

where

$$d_1(n) = \text{card} \{d \in \mathbb{N} : d \mid n, \ d \equiv 1 \pmod 4\} ,$$
$$d_3(n) = \text{card} \{d \in \mathbb{N} : d \mid n, \ d \equiv 3 \pmod 4\} .$$

Theorem 5.21 *We have*

$$r(n) = 4\delta(n) .$$

Proof Following the notation in (5.8) we write

$$n = 2^r \left(p_1^{r_1} \cdots p_k^{r_k}\right)\left(q_1^{s_1} \cdots q_\ell^{s_\ell}\right) . \qquad (5.13)$$

The odd divisors of n are precisely the monomials that we obtain by expanding

$$\prod_{j=1}^{k} \left(1 + p_j + p_j^2 + \ldots + p_j^{r_j}\right) \prod_{m=1}^{\ell} \left(1 + q_m + q_m^2 + \ldots + q_m^{s_m}\right) . \qquad (5.14)$$

The divisors of n which are $\equiv 1 \pmod 4$ are the monomials containing an even number of primes q counted with multiplicity, while the divisors $\equiv 3 \pmod 4$ are the monomials with an odd number of terms q. In order to compute $\delta(n)$ we therefore replace every p and q in (5.14) with $+1$ and -1, respectively. Then

$$\delta(n) = \begin{cases} \prod_{j=1}^{k} \left(r_j + 1\right) & \text{if the terms } s_m \text{ are all even,} \\ 0 & \text{otherwise.} \end{cases}$$

The conclusion now follows from Theorem 5.20. $\qquad \square$

We are now in a position to apply Theorem 2.13 and deduce similar estimates for the mean growth of $r(n)$.

Theorem 5.22 *The arithmetic function $r(n)$ satisfies the following conditions.*

(i) *For every $\varepsilon > 0$ there is a constant $c = c_\varepsilon$ such that*

$$r(n) \leq c\, n^\varepsilon .$$

(ii) *There are no positive constants c_1 and c_2 such that*

$$r(n) \leq c_1 \log^{c_2} n . \tag{5.15}$$

Proof (i) is a direct consequence of Theorem 5.21 and Theorem 2.13. In order to prove (ii), we assume that (5.15) is true and modify the argument in the proof of Theorem 2.13. By Corollary 3.10 there are infinitely many prime numbers $p_1 < p_2 < \ldots$ which are $\equiv 1 \pmod 4$. Choose an integer $h > c_2$ and let $a_n = (p_1 p_2 \cdots p_{h+1})^n$ for every positive integer n. Arguing as in the proof of Theorem 2.13, we show that $d(a_n) > K(c) \log^c(a_n)$, where $K(c) = \log^{-h-1}(p_1 p_2 \cdots p_{h+1})$ depends only on h. Observe that a_n has no divisors $\equiv 3 \pmod 4$, so that $\delta(a_n) = d_1(a_n) = d(a_n)$. □

5.4 Sums of four squares

We shall prove the following result, stated by Bachet in 1621 and proved[3] by Lagrange in 1770.

Theorem 5.23 (Lagrange) *Every positive integer can be written as a sum of four squares.*

Assume that a and b are sums of four squares. Then the same is true for ab. Indeed

$$\left(x_1^2 + x_2^2 + x_3^2 + x_4^2\right)\left(y_1^2 + y_2^2 + y_3^2 + y_4^2\right) \tag{5.16}$$
$$= (x_1 y_1 + x_2 y_2 + x_3 y_3 + x_4 y_4)^2 + (x_1 y_2 - x_2 y_1 + x_3 y_4 - x_4 y_3)^2$$
$$+ (x_1 y_3 - x_3 y_1 + x_4 y_2 - x_2 y_4)^2 + (x_1 y_4 - x_4 y_1 + x_2 y_3 - x_3 y_2)^2 .$$

Then Theorem 5.23 is a corollary of the following result.

Theorem 5.24 *Every prime number can be written as a sum of four squares.*

Let ω_d be the volume of the d-dimensional unit ball. Observe that

$$\omega_{d+1} = \omega_d \int_{-1}^{1} \left(\sqrt{1 - x^2}\right)^d dx . \tag{5.17}$$

---

[3] Lagrange acknowledged the influence of Euler. See [72, Vol. II, Ch. VIII] for a story of this result.

Then

$$\omega_4 = \omega_3 \int_{-1}^{1} \left(1 - x^2\right)^{3/2} dx = \frac{8}{3}\pi \int_{0}^{\pi/2} \cos^4 \theta \, d\theta$$

$$= \frac{1}{6}\pi \int_{0}^{\pi/2} \left(e^{i\theta} + e^{-i\theta}\right)^4 d\theta = \frac{1}{2}\pi^2 \ .$$

In general (see Lemma 11.1) we have $\omega_d = \frac{\pi^{d/2}}{\Gamma(\frac{d}{2}+1)}$, where

$$\Gamma(x) := \int_{0}^{+\infty} y^{x-1} e^{-y} \, dy \tag{5.18}$$

is the *gamma function*. Integration by parts proves the identities

$$\Gamma(x + 1) = x \, \Gamma(x) , \qquad \Gamma(n + 1) = n! \ . \tag{5.19}$$

Then the identity $\omega_{2k} = \pi^k/k!$ follows.

Lemma 5.25   *For every prime number p there exist $a, b \in \mathbb{Z}$ such that*

$$a^2 + b^2 \equiv -1 \pmod{p} \ . \tag{5.20}$$

Proof   Since $1^2 + 0^2 \equiv -1 \pmod 2$, we may assume that p is odd. Arguing as in the proof of Theorem 4.5 we observe that the $(p + 1)/2$ numbers in the set

$$A = \left\{ 0^2, 1^2, 2^2, \dots, \left(\frac{p-1}{2}\right)^2 \right\}$$

are pairwise non-congruent modulo p, and the same is true for the $(p + 1)/2$ numbers in the set

$$B = \left\{ -1 - 0^2, -1 - 1^2, -1 - 2^2, \dots, -1 - \left(\frac{p-1}{2}\right)^2 \right\} \ .$$

The set $A \cup B$ has $p + 1$ elements. Then there exist $a^2 \in A$ and $-1 - b^2 \in B$ such that $a^2 \equiv -1 - b^2 \pmod{p}$. $\qquad\square$

Proof of Theorem 5.24   Let L be the 4-dimensional lattice generated by

$$(1, 0, a, -b) , \quad (0, 1, b, a) , \quad (0, 0, p, 0) , \quad (0, 0, 0, p) ,$$

where a, b satisfy (5.20). Then $A(L)$ is (the absolute value of)

$$\det \begin{bmatrix} 1 & 0 & a & -b \\ 0 & 1 & b & a \\ 0 & 0 & p & 0 \\ 0 & 0 & 0 & p \end{bmatrix} ,$$

that is, $A(L) = p^2$. We take a ball $rB \subset \mathbb{R}^4$ centred at 0, with radius r and satisfying

$$\frac{2^{5/2}p}{\pi} < r^2 < 2p . \tag{5.21}$$

The volume of the 4-dimensional ball rB is equal to $\pi^2 r^4/2$. Then (5.21) yields

$$\frac{\pi^2}{2}r^4 > 2^4 p^2 = 2^4 A(L) .$$

By Minkowski's theorem, rB contains a point $0 \neq P \in L$. Let us write

$$P = \gamma_1 (1,0,a,-b) + \gamma_2 (0,1,b,a) + \gamma_3 (0,0,p,0) + \gamma_4 (0,0,0,p)$$
$$= (\gamma_1, \gamma_2, a\gamma_1 + b\gamma_2 + p\gamma_3, -b\gamma_1 + a\gamma_2 + p\gamma_4) .$$

Then

$$0 < \gamma_1^2 + \gamma_2^2 + (a\gamma_1 + b\gamma_2 + p\gamma_3)^2 + (-b\gamma_1 + a\gamma_2 + p\gamma_4)^2 < 2p . \tag{5.22}$$

Since

$$(a\gamma_1 + b\gamma_2)^2 + (-b\gamma_1 + a\gamma_2)^2 = (a^2 + b^2)(\gamma_1^2 + \gamma_2^2)$$

and since $1 + a^2 + b^2 \equiv 0 \pmod{p}$, we obtain

$$\gamma_1^2 + \gamma_2^2 + (a\gamma_1 + b\gamma_2 + p\gamma_3)^2 + (-b\gamma_1 + a\gamma_2 + p\gamma_4)^2 \tag{5.23}$$
$$\equiv \gamma_1^2 + \gamma_2^2 + (a\gamma_1 + b\gamma_2)^2 + (-b\gamma_1 + a\gamma_2)^2$$
$$\equiv (\gamma_1^2 + \gamma_2^2)(1 + a^2 + b^2) \equiv 0 \pmod{p} .$$

Then

$$\gamma_1^2 + \gamma_2^2 + (a\gamma_1 + b\gamma_2 + p\gamma_3)^2 + (-b\gamma_1 + a\gamma_2 + p\gamma_4)^2 = p .$$

In this way we have written p as a sum of four squares.   □

As an example, let us write 682 as a sum of four squares. Following (5.16) we write

$$682 = 22 \cdot 31$$
$$= \left(3^2 + 3^2 + 2^2 + 0^2\right)\left(5^2 + 2^2 + 1^2 + 1^2\right)$$
$$= (3 \cdot 5 + 3 \cdot 2 + 2 \cdot 1)^2 + (3 \cdot 2 - 3 \cdot 5 + 2 \cdot 1)^2$$
$$+ (3 \cdot 1 - 2 \cdot 5 - 3 \cdot 1)^2 + (3 \cdot 1 + 3 \cdot 1 - 2 \cdot 2)^2$$
$$= 23^2 + 7^2 + 10^2 + 2^2 .$$

Lagrange's theorem can be proved without appealing to Minkowski's theorem (see e.g. [9] or [63]).

Jacobi proved in 1828 the stronger result that the number of ways to write a positive integer n as a sum of four squares is eight times the sum of the divisors of n that are not multiples of 4 (see [7] or [128]).

If we want to write a positive integer as a sum of cubes, we need nine of them (not less, because $23 = 2 \cdot 2^3 + 7 \cdot 1^3$). Hilbert proved in 1909 the following general result (see e.g. [98]).

Theorem 5.26 (Hilbert) *For every positive integer k there exists a positive integer $g(k)$ such that every positive integer is a sum of $g(k)$ kth powers of non-negative integers.*

Exercises

1) Let k be a positive integer and let $A \subseteq \mathbb{R}^d$ be a measurable set with volume $> k$. Prove the existence of $k + 1$ different points $x_1, x_2, \ldots, x_{k+1}$ such that $x_i - x_j \in \mathbb{Z}^d$ for every i, j.

2) Let $D \subseteq \mathbb{R}^d$ be a convex set, symmetric around the origin. Assume that $D \cap \mathbb{Z}^d = \{0\}$. Prove that if p and q are different points in $\mathbb{Z}^d$, then

$$\left(\frac{1}{2}D + p\right) \cap \left(\frac{1}{2}D + q\right) = \varnothing .$$

3) Given a convex set $D \subseteq \mathbb{R}^d$, let ch (D) be the convex hull of D, that is, the intersection of all convex sets containing D. Prove that

$$\text{ch}\,(D) = \Big\{t \in \mathbb{R}^d : t = \alpha_1 t_1 + \ldots + \alpha_k t_k, \ k \in \mathbb{N}, \ t_1, \ldots, t_k \in D,$$
$$\alpha_1 \geq 0, \ldots, \alpha_k \geq 0, \ \alpha_1 + \ldots + \alpha_k = 1\Big\} .$$

4) Prove in an elementary way that no integer $m \equiv 3 \pmod{4}$ can be written as a sum of two squares.

5) Prove that q and r in Proposition 5.12 are not uniquely determined. Show that $r = 0$ if and only if $q \mid z$.

6) Let π be a Gaussian prime. Prove that there exists one and only one prime number p such that $\pi \mid p$.

7) Let (a, b, c) be a primitive Pythagorean triple. That is a, b, c are positive integers with no common factor and satisfying $a^2 + b^2 = c^2$. Prove that $a + ib$ and $a - ib$ are relatively prime in $\mathbb{Z}\,[i]$.

8) Let α, β be coprime Gaussian integers. Assume that $\alpha\beta$ is a square. Prove that there exists a unit u such that $u\alpha$ and $u^{-1}\beta$ are squares.

9) Give a proof of Lemma 5.19 without appealing to Gaussian integers.

10) Determine the smallest positive integer which can be written as a sum of two squares in more than eight ways.

11) Determine all representations of 325 as a sum of two squares.

12) Prove that a positive integer n can be written as a difference of two squares if and only if $n \not\equiv 2 \pmod 4$. Prove that this representation is unique (up to signs) when n is a prime number.

13) Write 874 as a sum of four squares.

PART TWO

FOURIER ANALYSIS AND GEOMETRIC DISCREPANCY

6

Uniform distribution and completeness of the trigonometric system

In Chapter 2 we have seen proofs of the 'elementary' estimates for the mean growth of the arithmetic functions $d(n)$ and $r(n)$ which consist of counting the integer points (that is, the points with integral coordinates) in certain domains in $\mathbb{R}^2$. In the second part of this book we shall use Fourier analytic methods to improve the above-mentioned elementary estimates and to study several related results concerning lattice points in large domains in $\mathbb{R}^d$. As an essentially more general point of view, we shall consider finite sequences $\left\{x_j\right\}_{j=1}^N \subset \mathbb{T}^d = \mathbb{R}^d/\mathbb{Z}^d$, and we shall investigate the 'discrepancy' between the integral $\int_{\mathbb{T}^d} f(x)\,dx$ and the Riemann sum $\frac{1}{N} \sum_{j=1}^N f\left(x_j\right)$, where f belongs to a suitable class of functions on $\mathbb{T}^d$. We begin the study of this multidimensional numerical integration problem by considering the case where f is the characteristic function of an arbitrary interval in $\mathbb{T}$ or $\mathbb{T}^d$. This will lead us to the definition of a *uniformly distributed sequence*, which is the main topic of this chapter.

Let us go back to Corollary 5.4, which says that for every irrational number α there exist infinitely many rational numbers p/j such that

$$\left| \alpha - \frac{p}{j} \right| < \frac{1}{j^2} \ .$$

Then for every interval $I_0 \subseteq \mathbb{T}$ containing the origin, and for every irrational number α, there exist infinitely many different points of the sequence $j\alpha$ which belong to I_0. We will show that the same is true for every interval $I \subseteq \mathbb{T}$, so that $j\alpha$ is dense in $\mathbb{T}$ (this means that the fractional parts[1] $\{j\alpha\} = j\alpha - [j\alpha]$ are

---

[1] To avoid any possible confusion, we shall always write $\{\beta\}$ for the fractional part of a real number β. A sequence will be written with no curly brackets, for example, $j\alpha$, $t(j)$, t_j or with brackets and explicit indices, for example, $\{j\alpha\}_{j=1}^\infty$, $\{t(j)\}_{j=1}^N$, $\{t_j\}_{j=1}^\infty$.

dense in $[0, 1)$). This comes as a consequence of Kronecker's theorem, a deep result that we are going to describe in the coming section.

6.1 Kronecker's theorem

The density of $j\alpha$ in $\mathbb{T}$ concerns the *set* $\{j\alpha\}_{j=1}^{\infty}$, while Kronecker's theorem deals with the *sequence* $\{j\alpha\}_{j=1}^{\infty}$, showing that, in a certain sense, this sequence fills all the intervals in $\mathbb{T}$ at a proper speed. More precisely, $j\alpha$ satisfies the following definition (see [86, 95, 114]).

Definition 6.1 Let $t(j)$ be a sequence taking values in $\mathbb{T}$. We say that $t(j)$ is *uniformly distributed in* $\mathbb{T}$ if for every interval $I \subseteq \mathbb{T}$ we have

$$\lim_{N \to \infty} \frac{\text{card}\{t(j) \in I : 1 \leq j \leq N\}}{N} = |I| ,$$

where $|I|$ is the length of I.

Observe that every interval in $[0, 1)$ is an interval in $\mathbb{T}$, while there are intervals in $\mathbb{T}$ which are only the union of two intervals in $[0, 1)$.

If $u(j)$ is a real sequence, we consider the sequence $t(j) = \{u(j)\}$ of its fractional parts. If $t(j)$ is uniformly distributed in $[0, 1)$, then we say that $u(j)$ is *uniformly distributed* mod 1. Sometimes we may simply say that $u(j)$ is uniformly distributed.

It is easy to extend Definition 6.1 to several variables. By an interval in $\mathbb{T}^d$ we shall mean the product $I = I_1 \times I_2 \times \ldots \times I_d$ of d intervals in $\mathbb{T}$.

Definition 6.2 A sequence $t(j)$ taking values in $\mathbb{T}^d$ is *uniformly distributed in* $\mathbb{T}^d$ if for every interval $I \subseteq \mathbb{T}^d$ we have

$$\lim_{N \to \infty} \frac{\text{card}\{t(j) \in I : 1 \leq j \leq N\}}{N} = |I| , \tag{6.1}$$

where $|I|$ is the volume of I.

We can now state Kronecker's theorem.

Theorem 6.3 (Kronecker) *Let the real numbers* $1, \alpha_1, \alpha_2, \ldots, \alpha_d$ *be linearly independent over* $\mathbb{Q}$. *Let* $\alpha = (\alpha_1, \alpha_2, \ldots, \alpha_d)$, *then the sequence* $j\alpha$ *is uniformly distributed in* $\mathbb{T}^d$.

We shall see the proof of this theorem in Section 6.3.

Recall that if we assume $1, \alpha_1, \alpha_2, \ldots, \alpha_d$ linearly independent over $\mathbb{Q}$, then the numbers $\alpha_1, \alpha_2, \ldots, \alpha_d$ must be irrational. The example $\alpha_1 = \sqrt{2}$, $\alpha_2 = \sqrt{2} - 1$ shows that the converse is not true when $d \geq 2$.

Going back to Definition 6.2, we rewrite (6.1) in the form

$$\lim_{N \to \infty} \left(\frac{1}{N} \sum_{j=1}^{N} \chi_I(t(j)) \right) = \int_{\mathbb{T}^d} \chi_I(t) \, dt \,, \tag{6.2}$$

which shows an integral equal to a limit of Riemann sums, at least for the class of characteristic functions[2] of intervals. Therefore the definition of the Riemann integral makes the following characterization natural.

Theorem 6.4 (Weyl) *A sequence $t(j)$ taking values in $\mathbb{T}^d$ is uniformly distributed if and only if*

$$\lim_{N \to \infty} \left(\frac{1}{N} \sum_{j=1}^{N} f(t(j)) \right) = \int_{\mathbb{T}^d} f(x) \, dx \tag{6.3}$$

for every Riemann integrable function f on $\mathbb{T}^d$.

Proof The 'if' part is obvious since (6.2) is a particular case of (6.3). Let us prove the 'only if' part. Let $t(j)$ be uniformly distributed in $\mathbb{T}^d$ and let f be a Riemann integrable function on $\mathbb{T}^d$. Then for every $\varepsilon > 0$ there exist two finite linear combinations

$$s(x) = \sum_{h} a_h \chi_{I_h}(x) \,, \qquad S(x) = \sum_{h} A_h \chi_{I_h}(x)$$

of characteristic functions of intervals I_h such that $s(x) \le f(x) \le S(x)$ for every x and $\int_{\mathbb{T}^d} (S(x) - s(x)) \, dx \le \varepsilon$. Since $t(j)$ is uniformly distributed, we have, for every h,

$$\lim_{N \to \infty} \left(\frac{1}{N} \sum_{j=1}^{N} \chi_{I_h}(t(j)) \right) = |I_h| \,.$$

Since the sum over h is finite, we have

$$\lim_{N \to \infty} \left(\frac{1}{N} \sum_{j=1}^{N} s(t(j)) \right) = \lim_{N \to \infty} \left(\frac{1}{N} \sum_{j=1}^{N} \sum_{h} a_h \chi_{I_h}(t(j)) \right)$$

$$= \sum_{h} a_h \lim_{N \to \infty} \left(\frac{1}{N} \sum_{j=1}^{N} \chi_{I_h}(t(j)) \right) = \sum_{h} a_h |I_h| = \int_{\mathbb{T}^d} s(x) \, dx \,,$$

---

[2] The characteristic function of a set E is $\chi_E(x) := \begin{cases} 1 & \text{if } x \in E, \\ 0 & \text{if } x \notin E. \end{cases}$

and the same holds true for $S(x)$. Hence the theorem follows from the definition of the Riemann integral and from the inequalities

$$\frac{1}{N} \sum_{j=1}^{N} s(t(j)) \leq \frac{1}{N} \sum_{j=1}^{N} f(t(j)) \leq \frac{1}{N} \sum_{j=1}^{N} S(t(j)) .$$

$\square$

The proof of Kronecker's theorem is an immediate corollary of a simple and deep characterization of uniformly distributed sequences, called the *Weyl criterion*, which is in turn related to the completeness of the trigonometric system.

6.2 Completeness of the trigonometric system

Let f be a measurable function on $\mathbb{T}^d$ and let $p \geq 1$ be a real number. We say that f belongs to $L^p\left(\mathbb{T}^d\right)$ if

$$\|f\|_{L^p(\mathbb{T}^d)} := \left\{ \int_{\mathbb{T}^d} |f(t)|^p \, dt \right\}^{1/p} < +\infty .$$

Observe that $\|f\|_{L^p(\mathbb{T}^d)} = 0$ if and only if $f(t) = 0$ for almost every $t \in \mathbb{T}^d$.

Let f be a measurable function on $\mathbb{T}^d$. We define $\|f\|_{L^\infty(\mathbb{T}^d)}$ to be the smallest K such that $|f(t)| \leq K$ for almost every $t \in \mathbb{T}^d$. Then $L^\infty\left(\mathbb{T}^d\right)$ is the space of all functions admitting such a K.

Theorem 6.5 (Hölder's inequality)   *Let $1 \leq p, q \leq \infty$ with $\frac{1}{p} + \frac{1}{q} = 1$ and let f, g be measurable non-negative functions on $\mathbb{T}^d$. Then*

$$\int_{\mathbb{T}^d} fg \leq \|f\|_{L^p(\mathbb{T}^d)} \|g\|_{L^q(\mathbb{T}^d)} . \tag{6.4}$$

Proof   If the two indices p, q take values 1 and ∞, then the inequality is simple. We therefore assume that $p > 1$ and $q > 1$. It is enough to prove (6.4) when $\|f\|_{L^p(\mathbb{T}^d)} = \|g\|_{L^q(\mathbb{T}^d)} = 1$. Moreover, without loss of generality, we may assume that f and g are positive on $\mathbb{T}^d$. For every $t \in \mathbb{T}^d$ let $s = s(t)$ and $u = u(t)$ be the real numbers satisfying $f(t) = e^{s/p}$ and $g(t) = e^{u/q}$. We may assume that $s \leq u$, and consequently observe that

$$s \leq s + \frac{u-s}{q} = \frac{s}{p} + \frac{u}{q} = u + \frac{s-u}{p} \leq u . \tag{6.5}$$

Let $\varphi(r) = ar + b$ be the straight line joining the points (s, e^s) and (u, e^u). By convexity we have $e^r \le \varphi(r)$ for every $s \le r \le u$. Then, by (6.5),

$$e^{\frac{s}{p}+\frac{u}{q}} \le \varphi\left(\frac{s}{p}+\frac{u}{q}\right) = a\left(\frac{s}{p}+\frac{u}{q}\right) + b = \frac{1}{p}(as+b) + \frac{1}{q}(au+b)$$

$$= \frac{1}{p}\varphi(s) + \frac{1}{q}\varphi(u) = \frac{1}{p}e^s + \frac{1}{q}e^u,$$

that is,

$$f(t)g(t) \le \frac{1}{p}f^p(t) + \frac{1}{q}g^q(t).$$

The integration of both sides gives

$$\int_{\mathbb{T}^d} fg \le \frac{1}{p}\int_{\mathbb{T}^d} f^p + \frac{1}{q}\int_{\mathbb{T}^d} g^q = 1.$$

$\square$

Corollary 6.6   *Let $1 \le p \le q \le +\infty$. Then*

$$\|f\|_{L^p(\mathbb{T}^d)} \le \|f\|_{L^q(\mathbb{T}^d)}.$$

Proof   We may assume $1 \le p < q < +\infty$, then $q/p > 1$. Since $\int_{\mathbb{T}^d} dt = 1$, Hölder's inequality implies

$$\|f\|_{L^p(\mathbb{T}^d)} = \left\{\int_{\mathbb{T}^d}|f|^p\right\}^{1/p} = \left\{\int_{\mathbb{T}^d}|f|^p \cdot 1\right\}^{1/p} \le \left\{\int_{\mathbb{T}^d}|f|^q\right\}^{1/q} \cdot 1 = \|f\|_{L^q(\mathbb{T}^d)}.$$

$\square$

Proposition 6.7 (Minkowski's inequality)   *For every $1 \le p \le \infty$ we have*

$$\|f+g\|_{L^p(\mathbb{T}^d)} \le \|f\|_{L^p(\mathbb{T}^d)} + \|g\|_{L^p(\mathbb{T}^d)}.$$

Proof   The result is obvious if $p = 1$ or $p = \infty$. Let $1 < p < \infty$. We may assume that $f \ne -g$. By Hölder's inequality we have

$$\int_{\mathbb{T}^d}|f+g|^p \le \int_{\mathbb{T}^d}(|f|+|g|)\,|f+g|^{p-1}$$

$$\le \|f\|_{L^p(\mathbb{T}^d)}\left\||f+g|^{p-1}\right\|_{L^q(\mathbb{T}^d)} + \|g\|_{L^p(\mathbb{T}^d)}\left\||f+g|^{p-1}\right\|_{L^q(\mathbb{T}^d)}$$

$$= \left(\|f\|_{L^p(\mathbb{T}^d)} + \|g\|_{L^p(\mathbb{T}^d)}\right)\left(\int_{\mathbb{T}^d}|f+g|^p\right)^{1-1/p},$$

since $\frac{1}{p}+\frac{1}{q} = 1$ means $q = p/(p-1)$. Then

$$\left\{\int_{\mathbb{T}^d}|f+g|^p\right\}^{1/p} \le \|f\|_{L^p(\mathbb{T}^d)} + \|g\|_{L^p(\mathbb{T}^d)}.$$

$\square$

Remark 6.8   Let $1 < p < \infty$ and let f be a real function belonging to $L^p\left(\mathbb{T}^d\right)$. Let

$$\widetilde{g}(t) = \begin{cases} |f(t)|^{p-1} & \text{if } f(t) \geq 0, \\ -|f(t)|^{p-1} & \text{if } f(t) < 0. \end{cases}$$

Then Hölder's inequality implies

$$\int_{\mathbb{T}^d} |f|^p = \int_{\mathbb{T}^d} f\widetilde{g} \leq \|f\|_{L^p(\mathbb{T}^d)} \|\widetilde{g}\|_{L^q(\mathbb{T}^d)}$$

$$= \left(\int_{\mathbb{T}^d} |f|^p\right)^{1/p} \left(\int_{\mathbb{T}^d} |f|^p\right)^{1/q} = \int_{\mathbb{T}^d} |f|^p \ .$$

Then for every non-zero function $f \in L^p\left(\mathbb{T}^d\right)$ there exists a non-zero function $\widetilde{g} \in L^q\left(\mathbb{T}^d\right)$ such that Hölder's inequality becomes an equality:

$$\int_{\mathbb{T}^d} f\widetilde{g} = \|f\|_{L^p(\mathbb{T}^d)} \|\widetilde{g}\|_{L^q(\mathbb{T}^d)} \ ,$$

that is,

$$\|f\|_{L^p(\mathbb{T}^d)} = \int_{\mathbb{T}^d} \frac{f\widetilde{g}}{\|\widetilde{g}\|_{L^q(\mathbb{T}^d)}} \ .$$

Whence

$$\|f\|_{L^p(\mathbb{T}^d)} = \sup_{\|g\|_{L^q(\mathbb{T}^d)}=1} \int_{\mathbb{T}^d} fg \ .$$

Theorem 6.9 (Minkowski's integral inequality)   *Let f be a measurable and non-negative function on $\mathbb{T}^d \times \mathbb{T}^d$, and let $1 \leq p < \infty$. Then*

$$\left\{\int_{\mathbb{T}^d} \left(\int_{\mathbb{T}^d} f(t,y)\, dy\right)^p dt\right\}^{1/p} \leq \int_{\mathbb{T}^d} \left\{\int_{\mathbb{T}^d} f^p(t,y)\, dt\right\}^{1/p} dy \ . \qquad (6.6)$$

This means that the L^p norm of the integral is not larger than the integral of the L^p norm.

Proof   When $p = 1$, (6.6) is a consequence of Tonelli's theorem. When $p > 1$, Remark 6.8 and Hölder's inequality imply

$$\left\{\int_{\mathbb{T}^d} \left(\int_{\mathbb{T}^d} f(t,y)\, dy\right)^p dt\right\}^{1/p} = \sup_{\|g\|_{L^q(\mathbb{T}^d)}=1} \left|\int_{\mathbb{T}^d} \left(\int_{\mathbb{T}^d} f(t,y)\, dy\right) g(t)\, dt\right|$$

$$= \sup_{\|g\|_{L^q(\mathbb{T}^d)}=1} \left|\int_{\mathbb{T}^d} \left(\int_{\mathbb{T}^d} f(t,y)g(t)\, dt\right) dy\right|$$

$$\leq \sup_{\|g\|_{L^q(\mathbb{T}^d)}=1} \int_{\mathbb{T}^d} \left(\left\{ \int_{\mathbb{T}^d} f^p(t,y)\, dt \right\}^{1/p} \left\{ \int_{\mathbb{T}^d} |g(t)|^q\, dt \right\}^{1/q} \right) dy$$

$$= \int_{\mathbb{T}^d} \left\{ \int_{\mathbb{T}^d} f^p(t,y)\, dt \right\}^{1/p} dy \ .$$

$\square$

For $f \in L^1(\mathbb{T}^d)$ and $n \in \mathbb{Z}^d$ we define the *Fourier coefficients*

$$\widehat{f}(n) := \int_{\mathbb{T}^d} f(t)\, e^{-2\pi i n \cdot t}\, dt \ ,$$

where $n \cdot t = n_1 t_1 + n_2 t_2 + \ldots + n_d t_d$ is the inner product in $\mathbb{R}^d$. If $f, g \in L^1(\mathbb{T}^d)$ we define the *convolution*

$$(f * g)(t) := \int_{\mathbb{T}^d} f(t-y) g(y)\, dy \ .$$

Proposition 6.10 *For every $f, g \in L^1(\mathbb{T}^d)$ we have*

(i) $(f * g)^\wedge (n) = \widehat{f}(n)\, \widehat{g}(n)$ *for every $n \in \mathbb{Z}^d$,*

(ii) $\|f * g\|_{L^p(\mathbb{T}^d)} \leq \|f\|_{L^p(\mathbb{T}^d)} \|g\|_{L^1(\mathbb{T}^d)}$ *if $f \in L^p\left(\mathbb{T}^d\right)$, $1 \leq p \leq \infty$.*

Proof   By the invariance of the Lebesgue measure under translation we have

$$(f * g)^\wedge (n) = \int_{\mathbb{T}^d} \left(\int_{\mathbb{T}^d} f(t-y) g(y) dy \right) e^{-2\pi i n \cdot t}\, dt$$

$$= \int_{\mathbb{T}^d} g(y) \int_{\mathbb{T}^d} f(t-y)\, e^{-2\pi i n \cdot t}\, dt dy$$

$$= \int_{\mathbb{T}^d} g(y)\, e^{-2\pi i n \cdot y} \int_{\mathbb{T}^d} f(u)\, e^{-2\pi i n \cdot u}\, du dy = \widehat{f}(n)\, \widehat{g}(n) \ .$$

By Minkowski's integral inequality we have

$$\|f * g\|_{L^p(\mathbb{T}^d)} = \left\{ \int_{\mathbb{T}^d} \left| \int_{\mathbb{T}^d} f(y-t) g(t)\, dt \right|^p dy \right\}^{1/p}$$

$$\leq \int_{\mathbb{T}^d} |g(t)| \left\{ \int_{\mathbb{T}^d} |f(y-t)|^p\, dy \right\}^{1/p} dt$$

$$= \int_{\mathbb{T}^d} |g(t)| \left\{ \int_{\mathbb{T}^d} |f(u)|^p\, du \right\}^{1/p} dt = \|g\|_{L^1(\mathbb{T}^d)} \|f\|_{L^p(\mathbb{T}^d)} \ .$$

$\square$

We call a *trigonometric polynomial on* $\mathbb{T}^d$ any finite sum of the form

$$P(t) = \sum a_n e^{2\pi i n \cdot t},$$

where $a_n \in \mathbb{C}$. When $d = 1$ we call the *degree* of P the largest $n \geq 0$ such that $a_n \neq 0$ or $a_{-n} \neq 0$.

By (i) in the previous proposition the convolution of a trigonometric polynomial of degree N with an integrable function is still a trigonometric polynomial of degree N.

As a very significant example, we introduce the *Fejér kernel* on $\mathbb{T}$. This is the family of trigonometric polynomials of degree N defined by

$$K_N(x) = K_N^o(x) := \sum_{j=-N}^{N} \left(1 - \frac{|j|}{N+1}\right) e^{2\pi i j x} \qquad (6.7)$$

(we shall write $K_N^o(x)$ only when it proves necessary to avoid confusion with the multidimensional Fejér kernel).

Lemma 6.11   *For every* $x \notin \mathbb{Z}$ *we have*

$$K_N(x) = \frac{1}{N+1} \left(\frac{\sin\left(\pi\left(N+1\right)x\right)}{\sin\left(\pi x\right)}\right)^2. \qquad (6.8)$$

Proof   By (4.27) we have, for every $x \notin \mathbb{Z}$,

$$\sum_{n=-N}^{N} e^{2\pi i n x} = \frac{\sin\left((2N+1)\pi x\right)}{\sin\left(\pi x\right)}. \qquad (6.9)$$

Therefore (6.9) implies

$$(N+1)\sin^2\left(\pi x\right) K_N(x) = (N+1)\sin^2\left(\pi x\right) \sum_{j=-N}^{N} \left(1 - \frac{|j|}{N+1}\right) e^{2\pi i j x} \qquad (6.10)$$

$$= \sin^2\left(\pi x\right) \sum_{j=-N}^{N} (N+1-|j|) e^{2\pi i j x} = \sin^2\left(\pi x\right) \sum_{\ell=0}^{N} \left\{\sum_{j=-\ell}^{\ell} e^{2\pi i j x}\right\}$$

$$= \sin^2\left(\pi x\right) \sum_{\ell=0}^{N} \frac{\sin\left(2\pi\left(\ell + \frac{1}{2}\right)x\right)}{\sin\left(\pi x\right)} = \sin\left(\pi x\right) \operatorname{Im}\left\{\sum_{\ell=0}^{N} e^{2\pi i\left(\ell + \frac{1}{2}\right)x}\right\}$$

$$= \sin\left(\pi x\right) \operatorname{Im}\left\{e^{\pi i x} \sum_{\ell=0}^{N} e^{2\pi i \ell x}\right\} = \sin\left(\pi x\right) \operatorname{Im}\left\{e^{\pi i x} \frac{e^{2\pi i(N+1)x} - 1}{e^{2\pi i x} - 1}\right\}$$

$$= \sin\left(\pi x\right) \operatorname{Im}\left\{e^{\pi i(N+1)x} \frac{\sin\left(\pi\left(N+1\right)x\right)}{\sin\left(\pi x\right)}\right\} = \sin^2\left(\pi\left(N+1\right)x\right),$$

where Im denotes the imaginary part. $\qquad \square$

The family of trigonometric polynomials

$$D_N(x) := \sum_{n=-N}^{N} e^{2\pi i n x} = \frac{\sin((2N+1)\pi x)}{\sin(\pi x)} \tag{6.11}$$

is called the *Dirichlet kernel* and it is important because the partial sum

$$S_N f(x) = \sum_{n=-N}^{N} \widehat{f}(n) e^{2\pi i n x}$$

(see Section 4.3) is the convolution of f with D_N :

$$S_N f = f * D_N .$$

Indeed

$$(f * D_N)^\wedge (n) = \widehat{f}(n) \, \widehat{D}_N(n) = \begin{cases} \widehat{f}(n) & \text{if } |n| \le N, \\ 0 & \text{if } |n| > N, \end{cases}$$

since $\widehat{D}_N(n) = 1$ when $|n| \le N$ and $\widehat{D}_N(n) = 0$ when $|n| > N$. Observe, see inside the chain of identities (6.10), that

$$K_N = \frac{1}{N+1} \sum_{j=0}^{N} D_j \tag{6.12}$$

is the arithmetic mean of the Dirichlet kernel.

Theorem 4.19 proved the existence of a continuous function with pointwise non-convergent Fourier series. We are going to see that in this sense the Fejér kernel behaves better than the Dirichlet kernel.

We call *Fejér means* the trigonometric polynomials

$$(K_N * f)(x) = \sum_{j=-N}^{N} \left(1 - \frac{|j|}{N+1}\right) \widehat{f}(n) e^{2\pi i j x} .$$

The Fejér kernel on $\mathbb{T}^d$ is defined as

$$K_N(t) := K_N^o(t_1) \cdot K_N^o(t_2) \cdot \ldots \cdot K_N^o(t_d) ,$$

where $t = (t_1, t_2, \ldots, t_d)$ and K_N^o is the 1-dimensional Fejér kernel (see (6.7)). The d-dimensional generalization of the property (6.12) leads to a different kind of multidimensional Fejér kernel (see [170]), which we are not going to discuss here.

Lemma 6.12 *The Fejér kernel on $\mathbb{T}^d$ satisfies the following conditions.*

(i) $K_N(t) \geq 0$ *for every* $N \in \mathbb{N}$ *and every* $t \in \mathbb{T}^d$. *Moreover*

$$1 = \widehat{K_N}(0) = \int_{\mathbb{T}^d} K_N(t) \, dt = \int_{\mathbb{T}^d} |K_N(t)| \, dt . \qquad (6.13)$$

(ii) *Let* $0 < \delta < \frac{1}{2}$ *and let* $B(0, \delta)$ *be the ball centred at the origin and having radius* δ. *Then*

$$\lim_{N \to +\infty} \int_{[-\frac{1}{2},\frac{1}{2}]^d \setminus B(0,\delta)} K_N(t) \, dt = 0 .$$

Proof Let K_N^o be the 1-dimensional Fejér kernel. Then (i) is a simple consequence of (6.8) and of the identity $\widehat{K_N^o}(0) = 1$. In order to prove (ii) we observe that, as $N \to \infty$, (6.8) yields

$$0 < \int_{|x| \geq \delta} K_N^o(x) \, dx \leq \frac{2}{N+1} \int_{\delta}^{1/2} \frac{1}{\sin^2(\pi x)} \, dx \leq \frac{c_\delta}{N+1} \longrightarrow 0 .$$

Then

$$\int_{[-\frac{1}{2},\frac{1}{2}]^d \setminus B(0,\delta)} K_N(t) \, dt \leq \int_{\left\{ t: \, \max_j |t_j| \geq \delta/\sqrt{d} \right\}} K_N(t) \, dt$$

$$\leq \sum_{j=1}^{d} \int_{|t_j| \geq \delta/\sqrt{d}} K_N^o(t_j) \, dt_j \left(\prod_{r \neq j} \int_{\mathbb{T}} K_N^o(t_r) \, dt_r \right)$$

$$= \sum_{j=1}^{d} \int_{|t_j| \geq \delta/\sqrt{d}} K_N^o(t_j) \, dt_j \longrightarrow 0 .$$

$\square$

Theorem 6.13 *Let* $f \in C\left(\mathbb{T}^d\right)$. *Then, as* $N \to \infty$,

$$\|K_N * f - f\|_{L^\infty(\mathbb{T}^d)} \to 0 ,$$

so that $(K_N * f)(t)$ *converges uniformly to* $f(t)$.
  Let $f \in L^p\left(\mathbb{T}^d\right)$, *with* $1 \leq p < \infty$. *Then, as* $N \to \infty$,

$$\|K_N * f - f\|_{L^p(\mathbb{T}^d)} \to 0 .$$

Proof Since $\int_{\mathbb{T}^d} K_N = 1$ we have, for every $f \in C\left(\mathbb{T}^d\right)$,

$$\sup_{t \in \mathbb{T}^d} |K_N * f(t) - f(t)| = \sup_{t \in \mathbb{T}^d} \left| \int_{\mathbb{T}^d} f(t-y) K_N(y) \, dy - f(t) \int_{\mathbb{T}^d} K_N(y) \, dy \right|$$

$$(6.14)$$

$$= \sup_{t \in \mathbb{T}^d} \left| \int_{\mathbb{T}^d} (f(t-y) - f(t)) K_N(y) \, dy \right| \leq \int_{\mathbb{T}^d} |K_N(y)| \sup_{t \in \mathbb{T}^d} |f(t-y) - f(t)| \, dy .$$

Choose $\varepsilon > 0$. Since f is uniformly continuous on $\mathbb{T}^d$, there exists $\delta > 0$ such that if $|y| < \delta$, then

$$\sup_{t \in \mathbb{T}^d} |f(t - y) - f(t)| < \varepsilon .$$

We may assume that f is not identically zero. Then by the previous lemma there exists N_0 such that, for $N \geq N_0$,

$$\int_{[-\frac{1}{2},\frac{1}{2}]^d \setminus B(0,\delta)} K_N(y) \, dy < \frac{\varepsilon}{2 \, \|f\|_{L^\infty(\mathbb{T}^d)}} .$$

We thus write

$$\int_{\mathbb{T}^d} |K_N(y)| \sup_{t \in \mathbb{T}^d} |f(t - y) - f(t)| \, dy = \int_{B(0,\delta)} + \int_{[-\frac{1}{2},\frac{1}{2}]^d \setminus B(0,\delta)} = E + F .$$

We observe that, by (6.13),

$$E \leq \varepsilon \int_{B(0,\delta)} |K_N(y)| \, dy \leq \varepsilon ,$$

while, for $N \geq N_0$,

$$F \leq 2 \|f\|_{L^\infty(\mathbb{T}^d)} \int_{[-\frac{1}{2},\frac{1}{2}]^d \setminus B(0,\delta)} |K_N(y)| \, dy \leq \varepsilon .$$

Then

$$\int_{\mathbb{T}^d} |K_N(y)| \sup_{t \in \mathbb{T}^d} |f(t - y) - f(t)| \, dy \leq 2\varepsilon ,$$

and the first part of the theorem is proved. In order to prove the second part let us recall that $C\left(\mathbb{T}^d\right)$ is dense in $L^p\left(\mathbb{T}^d\right)$ when $1 \leq p < \infty$. Then, for every $f \in L^p\left(\mathbb{T}^d\right)$, there exists $g \in C\left(\mathbb{T}^d\right)$ such that $\|f - g\|_{L^p(\mathbb{T}^d)} < \varepsilon$. The first part of this theorem tells us that we can find N such that

$$\|K_N * g - g\|_{L^p(\mathbb{T}^d)} \leq \|K_N * g - g\|_{L^\infty(\mathbb{T}^d)} \leq \varepsilon .$$

Finally, by (ii) in Proposition 6.10 we obtain

$$\|K_N * f - f\|_{L^p(\mathbb{T}^d)}$$
$$\leq \|K_N * f - K_N * g\|_{L^p(\mathbb{T}^d)} + \|K_N * g - g\|_{L^p(\mathbb{T}^d)} + \|g - f\|_{L^p(\mathbb{T}^d)}$$
$$\leq \|K_N\|_{L^1(\mathbb{T}^d)} \|f - g\|_{L^p(\mathbb{T}^d)} + \varepsilon + \varepsilon \leq 3\varepsilon .$$

$\square$

Since $K_N * f$ is a trigonometric polynomial, we immediately deduce the completeness of the trigonometric system, that is, the following result, which is a consequence of Theorems 6.13 and 4.15.

Corollary 6.14 *The space of the trigonometric polynomials is dense in* $C\left(\mathbb{T}^d\right)$ *and in* $L^p\left(\mathbb{T}^d\right)$ *(if* $1 \le p < \infty$*). In particular, it is dense in* $L^2\left(\mathbb{T}^d\right)$*. Then* $\left\{e^{2\pi i n \cdot t}\right\}_{n \in \mathbb{Z}^d}$ *is a complete orthonormal system and we obtain the following identity, proved by Parseval in 1799:*

$$\sum_{n \in \mathbb{Z}^d} \left|\widehat{f}(n)\right|^2 = \int_{\mathbb{T}^d} |f(t)|^2 \, dt \, . \tag{6.15}$$

We can now go back to the uniformly distributed sequences.

6.3 The Weyl criterion

The following theorem, proved by Weyl in 1916 [180, 181], constitutes the fundamental result of the theory of uniformly distributed sequences.

Theorem 6.15 (Weyl criterion) *A sequence* $t(j)$ *with values in* $\mathbb{T}^d$ *is uniformly distributed if and only if, for every* $0 \neq k \in \mathbb{Z}^d$*, we have*

$$\lim_{N \to \infty} \frac{1}{N} \sum_{j=1}^{N} e^{2\pi i k \cdot t(j)} = 0 \, . \tag{6.16}$$

Proof Let $t(j)$ be uniformly distributed. If we choose $f(y) = e^{2\pi i k \cdot y}$, with $k \neq 0$, and apply (6.3) we obtain

$$\lim_{N \to \infty} \frac{1}{N} \sum_{j=1}^{N} e^{2\pi i k \cdot t(j)} = \int_{\mathbb{T}^d} e^{2\pi i k \cdot y} \, dy = 0 \, .$$

The other direction depends on the results obtained in the previous section. Let $P(t) = \sum_k a_k e^{2\pi i k \cdot t}$ be a trigonometric polynomial on $\mathbb{T}^d$. Then, if we assume (6.16),

$$\frac{1}{N} \sum_{j=1}^{N} P(t(j)) = a_0 + \sum_{k \neq 0} a_k \left(\frac{1}{N} \sum_{j=1}^{N} e^{2\pi i k \cdot t(j)} \right) \longrightarrow a_0 = \int_{\mathbb{T}^d} P \tag{6.17}$$

as $N \to \infty$. Then (6.3) is true for the trigonometric polynomials. Now let $f \in C\left(\mathbb{T}^d\right)$ and $\varepsilon > 0$. By (6.14) we know that

$$\left| \int_{\mathbb{T}^d} (K_M * f(y)) \, dy - \int_{\mathbb{T}^d} f(y) \, dy \right| \le \int_{\mathbb{T}^d} |K_M * f(y) - f(y)| \, dy \tag{6.18}$$

$$\le \sup_{y \in \mathbb{T}^d} |(K_M * f)(y) - f(y)| < \varepsilon$$

if M is large enough. Since $K_M * f$ is a trigonometric polynomial, (6.17) implies

$$\left| \frac{1}{N} \sum_{j=1}^{N} (K_M * f)(t(j)) - \int_{\mathbb{T}^d} (K_M * f)(y)\, dy \right| \le \varepsilon$$

for N large enough. Then (6.18) yields

$$\left| \frac{1}{N} \sum_{j=1}^{N} f(t(j)) - \int_{\mathbb{T}^d} f(y)\, dy \right|$$

$$\le \frac{1}{N} \sum_{j=1}^{N} |f(t(j)) - (K_M * f)(t(j))|$$

$$+ \left| \frac{1}{N} \sum_{j=1}^{N} (K_M * f)(t(j)) - \int_{\mathbb{T}^d} (K_M * f)(y)\, dy \right|$$

$$+ \left| \int_{\mathbb{T}^d} (K_M * f(y))\, dy - \int_{\mathbb{T}^d} f(y)\, dy \right|$$

$$\le 3\varepsilon\,.$$

Then (6.3) is true for the continuous functions. Now let I be any interval in $\mathbb{T}^d$ and let $g_1, g_2 \in C\left(\mathbb{T}^d\right)$ satisfy $g_1(t) \le \chi_I(t) \le g_2(t)$ for every $t \in \mathbb{T}^d$ and $\int_{\mathbb{T}^d} (g_2 - g_1) \le \varepsilon$. Then

$$|I| - \varepsilon \le \int_{\mathbb{T}^d} g_2(t)\, dt - \varepsilon \le \int_{\mathbb{T}^d} g_1(t)\, dt = \lim_{N \to +\infty} \left(\frac{1}{N} \sum_{j=1}^{N} g_1(t(j)) \right)$$

$$\le \liminf_{N \to +\infty} \left(\frac{1}{N} \operatorname{card}\left(\{t(j)\}_{j=1}^{N}\right) \cap I \right) \le \limsup_{N \to +\infty} \left(\frac{1}{N} \operatorname{card}\left(\{t(j)\}_{j=1}^{N}\right) \cap I \right)$$

$$\le \lim_{N \to +\infty} \left(\frac{1}{N} \sum_{j=1}^{N} g_2(t(j)) \right) = \int_{\mathbb{T}^d} g_2(t)\, dt \le \int_{\mathbb{T}^d} g_1(t)\, dt + \varepsilon \le |I| + \varepsilon\,.$$

The proof is now complete. $\qquad\square$

We can now prove Kronecker's theorem.

Proof of Theorem 6.3 By the Weyl criterion it is enough to prove that for every $0 \ne k = (k_1, k_2, \ldots, k_d) \in \mathbb{Z}^d$ we have

$$\frac{1}{N} \sum_{j=1}^{N} e^{2\pi i j\alpha \cdot k} \longrightarrow 0$$

as $N \to \infty$. Indeed

$$\left| \frac{1}{N} \sum_{j=1}^{N} e^{2\pi i j \alpha \cdot k} \right| = \frac{1}{N} \left| \frac{e^{2\pi i (N+1)\alpha \cdot k} - 1}{e^{2\pi i \alpha \cdot k} - 1} - 1 \right| \leq \frac{1}{N} + \frac{2}{N} \frac{1}{\left| e^{2\pi i \alpha \cdot k} - 1 \right|} \longrightarrow 0 ,$$

since $1, \alpha_1, \alpha_2, \ldots, \alpha_d$ are linearly independent over $\mathbb{Q}$, and then we cannot have $k \neq 0$ and $\alpha \cdot k \in \mathbb{Z}$. $\qquad\square$

We shall now exhibit a few conditions which imply uniform distribution.

Proposition 6.16 *Let $t(j)$ be a uniformly distributed sequence in $\mathbb{T}^d$, let c be a constant and let $u(j)$ be a sequence in $\mathbb{T}^d$ such that $(t(j) - u(j)) \to c$. Then $u(j)$ is uniformly distributed.*

Proof By Definition 6.1 we may assume that $c = 0$. Let $0 \neq k \in \mathbb{Z}^d$ and let us write

$$\frac{1}{N} \sum_{j=1}^{N} e^{2\pi i k \cdot u(j)} = \frac{1}{N} \sum_{j=1}^{N} \left(e^{2\pi i k \cdot u(j)} - e^{2\pi i k \cdot t(j)} \right) + \frac{1}{N} \sum_{j=1}^{N} e^{2\pi i k \cdot t(j)} := A_N + B_N .$$

By the Weyl criterion we know that $B_N \to 0$ as $N \to \infty$. We have

$$|A_N| = \left| \frac{1}{N} \sum_{j=1}^{N} \left(e^{2\pi i k \cdot (u(j) - t(j))} - 1 \right) e^{2\pi i k \cdot t(j)} \right| \leq \frac{1}{N} \sum_{j=1}^{N} \left| e^{2\pi i k \cdot (u(j) - t(j))} - 1 \right|$$

$$\leq \frac{1}{N} \sum_{j=1}^{N} |2\pi i k \cdot (u(j) - t(j))| \leq \frac{2\pi |k|}{N} \sum_{j=1}^{N} |u(j) - t(j)| .$$

We have used the inequality

$$\left| e^{i\theta} - 1 \right| \leq |\theta| \qquad\qquad (6.19)$$

(that is, the fact that a chord is not longer than its arc) and the Cauchy–Schwarz inequality. Let $0 < \varepsilon < 1$ and choose M such that $|u(j) - t(j)| < \varepsilon$ if $j > M$. We may assume that $N \geq M \sqrt{d}/\varepsilon$. Since $|a - b| \leq \sqrt{d}/2$ for every $a, b \in \mathbb{T}^d$, we have

$$\frac{1}{N} \sum_{j=1}^{N} |u(j) - t(j)| = \frac{1}{N} \sum_{j=1}^{M} |u(j) - t(j)| + \frac{1}{N} \sum_{j=M+1}^{N} |u(j) - t(j)| \leq \varepsilon + \varepsilon .$$

By the Weyl criterion we deduce that $u(j)$ is a uniformly distributed sequence. $\qquad\square$

In order to find more uniformly distributed sequences we are going to present a result due to Fejér, which deals with sequences obtained by restrictions of functions defined on $[1, +\infty)$. Observe that Fejér's condition is geometric in nature, while Kronecker's theorem involves an arithmetic condition.

First of all we present the *Euler–Maclaurin summation formula*, proved independently by Euler and Maclaurin in 1735.

Lemma 6.17 (Euler–Maclaurin summation formula) *Let $g \in C^1([1, +\infty))$. Then for all integers $0 < M < N$ we have*

$$\sum_{h=M}^{N} g(h) = \frac{1}{2}(g(M) + g(N)) + \int_M^N g(x)dx + \int_M^N s(x)g'(x)dx ,$$

where $s(x)$ is the sawtooth function (see (4.33)).

Proof Integrating by parts and recalling that $s(\ell^\pm) = \mp 1/2$ for every $\ell \in \mathbb{Z}$, and that $s'(x) = 1$ for every $x \notin \mathbb{Z}$, we have

$$\int_M^N s(x)g'(x)\,dx = \sum_{h=M}^{N-1} \int_h^{h+1} s(x)g'(x)\,dx$$

$$= \sum_{h=M}^{N-1} [s(x)g(x)]_{x=h^+}^{x=(h+1)^-} - \sum_{h=M}^{N-1} \int_h^{h+1} g(x)\,dx$$

$$= \sum_{h=M}^{N-1} \left(\frac{1}{2}g(h+1) + \frac{1}{2}g(h)\right) - \int_M^N g(x)\,dx$$

$$= \sum_{h=M}^{N} g(h) - \frac{1}{2}(g(N) + g(M)) - \int_M^N g(x)\,dx .$$

$\square$

Theorem 6.18 (Fejér) *Let $f \in C^2([1, +\infty))$ have $f''(x)$ definitely of constant sign. Assume that*

$$\lim_{x\to+\infty} \frac{1}{xf'(x)} = \lim_{x\to+\infty} \frac{f(x)}{x} = 0. \tag{6.20}$$

Then the sequence $f(j)$ is uniformly distributed in $\mathbb{T}$.

Proof By our assumptions we know that $f'(x)$ is monotonic for large x. Then there exists x_0 such that $|f''(x)|$ and $|f'(x)|$ are positive for every $x \geq x_0$. By the Weyl criterion it is enough to prove that for every integer $k \neq 0$ we have

$$\frac{1}{N} \sum_{j=1}^{N} e^{2\pi i k f(j)} \longrightarrow 0$$

as $N \to \infty$. Indeed, by the Euler–Maclaurin summation formula we have

$$\frac{1}{N} \sum_{j=1}^{N} e^{2\pi i k f(j)}$$

$$= \frac{1}{N} \int_{1}^{N} e^{2\pi i k f(x)} dx + \frac{1}{2N} \left(e^{2\pi i k f(1)} + e^{2\pi i k f(N)} \right)$$

$$+ \frac{1}{N} \int_{1}^{N} s(x) 2\pi i k f'(x) e^{2\pi i k f(x)} dx$$

$$:= A_N + B_N + C_N .$$

Of course $B_N \to 0$. We shall first of all show that $A_N \to 0$. Indeed, integrating by parts and recalling the assumptions in (6.20) we have, as $N \to \infty$,

$$|A_N| = \frac{1}{N} \left| \int_{1}^{N} e^{2\pi i k f(x)} dx \right| = o(1) + \frac{1}{N} \left| \int_{x_0}^{N} e^{2\pi i k f(x)} dx \right|$$

$$= o(1) + \frac{1}{N} \left| \int_{x_0}^{N} \left(e^{2\pi i k f(x)} 2\pi i k f'(x) \right) \frac{1}{2\pi i k f'(x)} dx \right|$$

$$\leq o(1) + \frac{1}{N} \left| \frac{e^{2\pi i k f(N)}}{2\pi i k f'(N)} \right| + \frac{1}{2\pi |k| N} \left| \int_{x_0}^{N} e^{2\pi i k f(x)} \frac{f''(x)}{[f'(x)]^2} dx \right|$$

$$\leq o(1) + \frac{1}{2\pi |k| N} \left| \int_{x_0}^{N} \frac{f''(x)}{[f'(x)]^2} dx \right| = o(1) + \frac{1}{2\pi |k| N |f'(N)|} \to 0$$

and

$$|C_N| = \frac{1}{N} \left| \int_{1}^{N} s(x) 2\pi i k f'(x) e^{2\pi i k f(x)} dx \right| \leq o(1) + \frac{1}{N} 2\pi |k| \left| \int_{x_0}^{N} f'(x) dx \right|$$

$$= o(1) + \frac{2\pi |k f(N)|}{N} \to 0 .$$

$\square$

As a consequence we immediately obtain the following result.

Corollary 6.19 *The sequences*

$$j^\alpha \quad (0 < \alpha < 1) \quad and \quad \log^\beta j \quad (\beta > 1)$$

are uniformly distributed mod 1.

Remark 6.20 The sequence $\log j$ is not uniformly distributed mod 1. Indeed, by Lemma 6.17,

$$\frac{1}{N} \sum_{j=1}^{N} e^{2\pi i \log j} = o(1) + \frac{1}{N} \int_{1}^{N} e^{2\pi i \log x} dx + \frac{1}{N} \int_{1}^{N} s(x) e^{2\pi i \log x} \frac{2\pi i}{x} dx$$

$$= o(1) + \frac{1}{N} \int_0^{\log N} e^{(2\pi i + 1)y} \, dy + O\left(\frac{\log N}{N}\right)$$

$$= o(1) + \frac{1}{N} \left(\frac{e^{(2\pi i + 1)\log N} - 1}{2\pi i + 1}\right) = o(1) + \frac{e^{2\pi i \log N} - 1/N}{2\pi i + 1} \nrightarrow 0 .$$

Note that $\log j$ is dense in $\mathbb{T}$ since

$$\log j \to +\infty , \quad (\log(j+1) - \log j) \longrightarrow 0 .$$

6.4 Normal numbers

By Kronecker's theorem we know that the sequence ne is uniformly distributed mod 1. Of course this cannot be true for every subsequence of ne. An explicit example is to consider the sequence $n!e$ and show that the fractional parts $\{n!e\}$ tend to zero, so that they are not uniformly distributed in $\mathbb{T}$. Indeed, as $n \to \infty$,

$$\{n!e\} = \left\{ n! \sum_{j=0}^{+\infty} \frac{1}{j!} \right\} = \left\{ n! \left(\sum_{j=0}^{n} \frac{1}{j!} + \sum_{j=n+1}^{+\infty} \frac{1}{j!} \right) \right\} = \left\{ n! \sum_{j=n+1}^{+\infty} \frac{1}{j!} \right\}$$

$$= \left\{ \frac{n!}{(n+1)!} + \frac{n!}{(n+2)!} + \frac{n!}{(n+3)!} + \cdots \right\}$$

$$= \left\{ \frac{1}{n+1} + \frac{1}{(n+1)(n+2)} + \frac{1}{(n+1)(n+2)(n+3)} + \cdots \right\}$$

$$= \left\{ \frac{1}{n+1} \left(1 + \frac{1}{n+2} + \frac{1}{(n+2)(n+3)} + \cdots \right) \right\} \sim \left\{ \frac{c}{n+1} \right\} \longrightarrow 0 .$$

The following result, proved by Weyl in 1916 [181], shows that the previous example (the fact that $n!e$ is not uniformly distributed) is, in a certain sense, an exception.

Theorem 6.21 *Let a_n be a sequence of pairwise distinct integers. Then the sequence $a_n x$ is uniformly distributed* mod 1 *for almost every real number x.*

Proof For every integer $k \neq 0$ let

$$P_N(x) := \frac{1}{N} \sum_{n=1}^{N} e^{2\pi i k a_n x} .$$

By our assumptions the terms a_n are pairwise distinct, and the same holds true for the numbers $k a_n$. Therefore the Parseval identity (6.15) implies

$$\int_0^1 |P_N(x)|^2 \, dx = \frac{1}{N^2} \int_0^1 \left| \sum_{n=1}^{N} e^{2\pi i k a_n x} \right|^2 dx = \frac{1}{N^2} \sum_{n=1}^{N} 1 = \frac{1}{N} .$$

If $N = m^2$ we have

$$\int_0^1 \sum_{m=1}^{+\infty} |P_{m^2}(x)|^2 \, dx = \sum_{m=1}^{+\infty} \int_0^1 |P_{m^2}(x)|^2 \, dx = \sum_{m=1}^{+\infty} \frac{1}{m^2} < \infty .$$

Hence $\sum_{m=1}^{+\infty} |P_{m^2}(x)|^2 < \infty$ for almost every x, and then $\lim_{m\to+\infty} P_{m^2}(x) = 0$ for almost every x. In order to end the proof we show that $\lim_{N\to+\infty} P_N(x) = 0$ for almost every x, and apply the Weyl criterion. Indeed, for every positive integer N let m be the positive integer satisfying $m^2 \le N < (m + 1)^2$. Then

$$|P_N(x)| = \left| \frac{1}{N} \sum_{n=1}^{N} e^{2\pi i k a_n x} \right| \le \left| \frac{1}{m^2} \sum_{n=1}^{m^2} e^{2\pi i k a_n x} \right| + \frac{1}{m^2} \sum_{n=m^2+1}^{m^2+2m} 1 = |P_{m^2}(x)| + \frac{2}{m} ,$$

which shows that $P_N(x) \to 0$ for almost every x.   □

Theorem 6.21 implies a classical result concerning real numbers.

Definition 6.22   Let $\alpha \in \mathbb{R}$ and let B_k be a sequence of k digits. Let us write $A_N(B_k)$ for the number of blocks of digits equal to B_k (possibly partially over-lapped) appearing among the first N digits of α. Assume that for every k and for every choice of B_k (note that there are 10^k such choices) we have

$$\lim_{N\to\infty} \frac{A_N(B_k)}{N} = 10^{-k} .$$

Then we shall say that α is normal.

If α is normal then a telephone number of 11 digits appears asymptotically every 10^{11} digits, within the digit expansion of α (10^{-11} is the theoretical prob-ability that a block of 11 digits equals the above telephone number). Observe that no rational number is normal, and that it is easy to construct irrational num-bers which are not normal (simply take an irrational number where the digit 6 never appears). Anyway, normality is the rule, as a famous result proved by Borel in 1909 and Cantelli in 1917 shows [21, 39, 54, 114, 146].

Theorem 6.23 (Borel–Cantelli)   *Almost every real number is normal.*

The proof of this theorem is an immediate consequence of Theorem 6.21 and of the following lemma, proved by Wall [179] in 1949. Wall's result re-lates the uniformly distributed sequences to the normal numbers, that is, to the strong law of large numbers (see Remark 6.26 below). It is interesting to re-mark that this connection was pointed out more than 30 years after the almost simultaneous appearance of the results of Weyl and Borel–Cantelli.

Lemma 6.24 (Wall)   *A real number α is normal if and only if the sequence $10^n \alpha$ is uniformly distributed* mod 1.

Proof We may assume that $\alpha > 0$. We write the decimal expansion $\alpha = [\alpha].a_1 a_2 \ldots$ Consider a block $B_k = b_1 b_2 \ldots b_k$ of k digits and assume that $a_{m+1} a_{m+2} \ldots a_{m+k} = b_1 b_2 \ldots b_k$. This means

$$\alpha = [\alpha] + \sum_{n=1}^{m} \frac{a_n}{10^n} + \left(\frac{b_1}{10^{m+1}} + \frac{b_2}{10^{m+2}} + \ldots + \frac{b_k}{10^{m+k}} \right) + \sum_{n=m+k+1}^{+\infty} \frac{a_n}{10^n},$$

that is,

$$\{10^m \alpha\} = \frac{b_1}{10} + \frac{b_2}{10^2} + \ldots + \frac{b_k}{10^k} + \sum_{j=k+1}^{+\infty} \frac{a_{j+m}}{10^j}.$$

This means that the fractional part $\{10^m \alpha\}$ belongs to the interval

$$I_k := \left[\frac{b_1}{10} + \frac{b_2}{10^2} + \ldots + \frac{b_k}{10^k} , \frac{b_1}{10} + \frac{b_2}{10^2} + \ldots + \frac{b_k}{10^k} + \frac{1}{10^k} \right].$$

Then the number $A_N(B_k)$ of blocks equal to B_k appearing among the first N digits of α is equal to the number of elements of the finite sequence $\{\{10^m \alpha\}\}_{m+k \leq N}$ which belong to I_k,

$$A_N(B_k) := \sum_{\substack{1 \leq m \leq N-k \\ \{10^m \alpha\} \in I_k}} 1. \tag{6.21}$$

Let $\{10^m \alpha\}_{m=1}^{\infty}$ be uniformly distributed mod 1. Then, by (6.21),

$$\lim_{N \to \infty} \frac{1}{N} A_N(B_k) = \lim_{N \to \infty} \frac{1}{N} \sum_{\substack{1 \leq m \leq N-k \\ \{10^m \alpha\} \in I_k}} 1$$

$$= \lim_{N \to \infty} \frac{N-k}{N} \left(\frac{1}{N-k} \sum_{\substack{1 \leq m \leq N-k \\ \{10^m \alpha\} \in I_k}} 1 \right) = |I_k| = 10^{-k}.$$

Hence α is normal and the 'if' part of the lemma is proved. Now let α be normal. Then

$$|I_k| = 10^{-k} = \lim_{N \to \infty} \frac{1}{N} A_N(B_k) = \lim_{N \to \infty} \frac{1}{N-k} \sum_{\substack{1 \leq m \leq N-k \\ \{10^m \alpha\} \in I_k}} 1$$

and if we use all the possible choices of B_k we shall see that the sequence $\{10^m \alpha\}_{m=1}^{\infty}$ satisfies Definition 6.2 for those intervals whose extremes are rational numbers with powers of 10 as denominators (that is, rational numbers with finite-digit expansion). We use the density of these numbers to end the proof. Indeed, let $[\alpha, \beta]$ be any interval in $\mathbb{T}$. Let a_1, a_2, b_1, b_2 have finite-digit sequences and satisfy $[a_1, b_1] \subset [\alpha, \beta] \subset [a_2, b_2]$ and $(b_2 - b_1) + (a_2 - a_1) < \varepsilon$. Then

$$\frac{1}{N} \text{ card } \{t(j) \in [a_1, b_1] : 1 \le j \le N\} - (b_1 - a_1) - \varepsilon$$

$$< \frac{1}{N} \text{ card } \{t(j) \in [\alpha, \beta] : 1 \le j \le N\} - (\beta - \alpha)$$

$$< \frac{1}{N} \text{ card } \{t(j) \in [a_2, b_2] : 1 \le j \le N\} - (b_2 - a_2) + \varepsilon,$$

and the lemma is proved. $\qquad \square$

Remark 6.25 It is quite difficult to construct normal numbers. The first example was provided by Champernowne [41]:

$$0.123456789101112131415\cdots$$

Champernowne conjectured that also

$$0.2357111317192329\cdots$$

(obtained by writing the sequence of prime numbers) is normal. Copeland and Erdős [62] proved the following more general fact. Let p_j be a strictly increasing sequence of positive integers such that

$$\text{card}\left\{p_j : j \le N\right\} \ge cN^{1-\varepsilon}$$

(observe that the prime numbers satisfy this density condition because of Theorem 1.15). Then the number $0.p_1 p_2 p_3 \cdots$ is normal.

Remark 6.26 Normal numbers have been studied by Borel and Cantelli while dealing with the *strong law of large numbers* (see e.g. [146]). Suppose we flip a coin infinitely many times, and add 0 for each tail and 1 for each head. Let S_N be the sum obtained after N tosses. The strong law of large numbers says that almost surely we have $\lim_{N \to +\infty} \frac{S_N}{N} = \frac{1}{2}$. In other words, we may see the above sequence of 0 and 1 as a number belonging to the interval $[0, 1)$, written in base 2, then the law says that almost every number in the interval $[0, 1)$ has asymptotically the same number of 0s and 1s. This is Theorem 6.23 for base 2 and blocks of one digit. The popular version of this result says that a monkey typing at random on a typewriter keyboard for an infinite amount of time will almost surely produce Dante's *Divine Comedy*. We actually know more than this: the monkey will produce it infinitely many times, with due frequency. We also know that we should not be in a hurry.

6.5 Benford's law

In a short note [129] published in 1881, Newcomb wrote:

That the ten digits do not occur with equal frequency must be evident to anyone making use of logarithmic tables, and noticing how much faster the first pages wear out than the last ones. The first significant digit is oftener 1 than any other digit, and the frequency diminishes up to 9 ... The law of probability of the occurrence of numbers is such that all mantissæ of their logarithms are equally probable.

The *first significant digit* is the first non-zero digit appearing in the digit expansion of a positive number (we agree to write, say, 0.5 and not $0.4999\cdots$). For example, the first significant digits of the numbers $\pi = 3.14159265\cdots$, 2014 and $1/2014 = 0.00049652\cdots$ are, respectively, 3, 2 and 4. Observe that a positive real number v has first significant digit k if and only if

$$\log_{10}(k) \le \{\log_{10}(v)\} < \log_{10}(k+1) , \tag{6.22}$$

where $\{\log_{10}(v)\}$ is the fractional part of $\log_{10}(v)$. In his paper Newcomb stated that the probability of the first significant digit being k is equal to the length

$$\log_{10}(k+1) - \log_{10}(k) = \log_{10}(1 + 1/k)$$

of the interval $[\log_{10}(k), \log_{10}(k+1))$. Let us write the values of these lengths for $k = 1, 2, \ldots, 9$:

$$
\begin{aligned}
&\log_{10}(2/1) = 0.30103\cdots && \log_{10}(3/2) = 0.17609\cdots \\
&\log_{10}(4/3) = 0.12494\cdots && \log_{10}(5/4) = 0.09691\cdots \\
&\log_{10}(6/5) = 0.079181\cdots && \log_{10}(7/6) = 0.066947\cdots \\
&\log_{10}(8/7) = 0.057992\cdots && \log_{10}(9/8) = 0.051153\cdots \\
&\log_{10}(10/9) = 0.045757\cdots
\end{aligned}
\tag{6.23}
$$

Therefore the first significant digit should be 1 with probability about 30.1%, 2 with probability about 17.6%, ... , 9 with probability about 4.6%.

Newcomb gave no actual numerical data or evidence for this 'logarithmic law', but in a sense he was not wrong. Indeed, in 1938 Benford [17], unaware of Newcomb's paper, produced many sequences (areas of rivers, populations, addresses, powers of integers, factorials, etc.) taken from the 'real world' and from mathematics, which showed, especially if taken as a whole, a good verification of the above-mentioned 'logarithmic law', which was later named after him.[3]

Let us check Benford's law on the populations of the Italian cities and towns (we say towns for all of them). There are 8092 towns in Italy, and for every

---

[3] Diaconis and Freeman have shown in [70] that Benford's data were manipulated, but even the unmanipulated data still show a remarkable evidence of the law.

$k = 1, 2, \ldots, 9$ we shall write $C(k)$ for the number of towns with population starting with the digit k:

$$C(1) = 2510 \approx 31.010\% \qquad C(2) = 1400 \approx 17.297\%$$
$$C(3) = 1013 \approx 12.515\% \qquad C(4) = 734 \approx 9.068\%$$
$$C(5) = 637 \approx 7.870\% \qquad C(6) = 545 \approx 6.733\%$$
$$C(7) = 461 \approx 5.695\% \qquad C(8) = 444 \approx 5.485\%$$
$$C(9) = 348 \approx 4.299\%$$

($2510 \approx 31.010\%$ means that 2510 is about 31.010% of 8092, ...). Note that these percentages fit very well with the numbers $\log_{10}(1 + 1/k)$ in (6.23).

Up to now we have not described a mathematical theorem, but rather a phenomenon of the real world. It is clear that Benford's law can be applied only to large data sets which range over several orders of magnitude. Anyway, as pointed out in [18], a satisfactory explanation of the law does not seem to be at hand.

Let us turn for the moment to a different feature of Benford's law and show that some familiar numerical sequences such as 2^n or $n!$ satisfy the law, which it is now time to state.

Definition 6.27   A positive real sequence t_n is a *Benford sequence* if, for every $k \in \{1, 2, \ldots, 9\}$,

$$\lim_{N \to +\infty} \frac{\text{card}\,\{n \le N : k \text{ is the first significant digit of } t_n\}}{N} = \log_{10}\left(1 + \frac{1}{k}\right).$$

A modification of the definition above will quickly relate Benford's law to the main topic of this chapter.

Definition 6.28   Let v be a positive real number and let $M(v)$ be the integer satisfying $1 \le 10^{-M(v)}v < 10$. Let $r \in \mathbb{N}$ and let $u_1 \in \{1, 2, \ldots, 9\}$, $u_\ell \in \{0, 1, \ldots, 9\}$ for every $\ell = 2, 3, \ldots, 9$. If $10^{-M(v)}v$ has digital expansion

$$10^{-M(v)}v = u_1.u_2 u_3 \ldots u_r \ldots$$

we say that the number $u_1 u_2 \ldots u_r$ is the *first significant r-block of v*.

Definition 6.29   A positive real sequence t_n is a *strong Benford sequence* if for every $r \in \mathbb{N}$ and every finite sequence $\{u_j\}_{j=1}^{r}$ such that $u_1 \in \{1, 2, \ldots, 9\}$ and $u_\ell \in \{0, 1, 2, \ldots, 9\}$ for every $\ell = 2, 3, \ldots, 9$, we have

$$\lim_{N \to +\infty} \frac{\text{card}\,\{n \le N : u_1 u_2 \ldots u_r \text{ is the first significant } r\text{-block of } t_n\}}{N}$$

$$= \log_{10}\left(1 + \frac{1}{u_1 u_2 \dots u_r}\right).$$

Of course, strong Benford sequences are Benford sequences.

The following result was proved by Diaconis in 1977 [69].

Theorem 6.30 (Diaconis)   *A real positive sequence t_n is a strong Benford sequence if and only if the sequence $\log_{10}(t_n)$ is uniformly distributed mod 1.*

In order to prove the 'if' part, recall (6.22) and apply Definition 6.1 to the intervals

$$\left[\log_{10}\left(u_1 + \frac{u_2}{10} + \dots + \frac{u_r}{10^{r-1}}\right), \log_{10}\left(u_1 + \frac{u_2}{10} + \dots + \frac{u_r}{10^{r-1}} + \frac{1}{10^r}\right)\right].$$

The proof of the 'only if' part is similar to the last step of the proof of Lemma 6.24 (see also [69] or [134, p. 111]).

Corollary 6.31   *The sequence 2^n is a strong Benford sequence.*

Proof   We need to show that $\log_{10}(2^n) = n\log_{10}(2)$ is uniformly distributed mod 1. This follows from Kronecker's theorem since $\log_{10}(2)$ is an irrational number.   □

Stirling's formula (see e.g. [150]) will allow us to prove that $n!$ is also a strong Benford sequence.

Theorem 6.32 (Stirling's formula)   *We have*

$$n! \sim \sqrt{2\pi n}\, n^n\, e^{-n}.$$

Proof   We know that $n! = \int_0^{+\infty} t^n e^{-t}\, dt$. Then, by the change of variables $t = n + s\sqrt{2n}$ we have

$$n! = \int_{-\sqrt{n/2}}^{+\infty} \left(n + s\sqrt{2n}\right)^n e^{-n - s\sqrt{2n}} \sqrt{2n}\, ds$$

$$= n^{n+1/2} e^{-n} \sqrt{2} \int_{-\sqrt{n/2}}^{+\infty} \left(1 + s\sqrt{2/n}\right)^n e^{-s\sqrt{2n}}\, ds$$

$$:= n^{n+1/2} e^{-n} \sqrt{2} \int_{-\infty}^{+\infty} g_n(s)\, ds,$$

where

$$g_n(s) = \begin{cases} 0 & \text{if } s \le -\sqrt{n/2}, \\ \left(1 + s\sqrt{2/n}\right)^n e^{-s\sqrt{2n}} = e^{n\log\left(1 + s\sqrt{2/n}\right) - s\sqrt{2n}} & \text{if } s > -\sqrt{n/2}. \end{cases}$$

We claim that

$$\lim_{n \to +\infty} \int_{-\infty}^{+\infty} g_n(s)\, ds = \int_{-\infty}^{+\infty} e^{-s^2}\, ds = \sqrt{\pi}. \tag{6.24}$$

The second equality in (6.24) is well known:

$$\left(\int_{-\infty}^{+\infty} e^{-s^2}\,ds\right)^2 = \int_{-\infty}^{+\infty} e^{-s^2}\,ds \int_{-\infty}^{+\infty} e^{-x^2}\,dx = \int_{\mathbb{R}^2} e^{-s^2-x^2}\,ds\,dx \qquad (6.25)$$

$$= \int_0^{2\pi}\int_0^{+\infty} e^{-\rho^2}\,\rho\,d\rho\,d\theta = \pi \int_0^{+\infty} e^{-v}\,dv = \pi\,.$$

Observe that

$$n\log\left(1 + s\,\sqrt{2/n}\right) - s\,\sqrt{2n} = n\left(\frac{s\sqrt{2}}{\sqrt{n}} - \frac{s^2}{n} + O\left(\frac{s^3}{n^{3/2}}\right)\right) - s\,\sqrt{2n}$$

$$= -s^2 + O\left(\frac{s^3}{\sqrt{n}}\right)$$

as $n \to +\infty$. In order to prove (6.24), let

$$k(x) = \frac{\log(1+x) - x}{x^2}$$

for $x > -1$. Then

$$k'(x) = \frac{\frac{x}{1+x} + x - 2\log(1+x)}{x^3} := \frac{a(x)}{x^3}\,.$$

Since $a(x) < 0$ for $-1 < x < 0$ and $a(x) > 0$ for $x > 0$, we see that $k(x)$ is increasing on $(-1, +\infty)$. The function $k(x)$ becomes continuous if we set $k(0) = -1/2$. Then, for every $n \in \mathbb{N}$,

$$e^{n\left(\log\left(1+s\sqrt{2/n}\right)-s\sqrt{2/n}\right)} = e^{2s^2\,k\left(s\sqrt{2/n}\right)}$$

$$\le \begin{cases} e^{2s^2\,k(0)} = e^{-s^2} & \text{if } -\sqrt{n/2} < s < 0, \\ e^{2s^2k\left(s\sqrt{2}\right)} = e^{-s\sqrt{2}}\left(1 + s\sqrt{2}\right) & \text{if } s > 0. \end{cases}$$

Hence the proof of the first equality in (6.24) follows from the dominated convergence theorem. □

Theorem 6.33 *n! is a strong Benford sequence.*

Proof By Stirling's formula we have

$$\log_{10}(n!) - \log_{10}\left(\sqrt{2\pi n}\,n^n e^{-n}\right) \longrightarrow 0\,,$$

and by Proposition 6.16 it is enough to prove that $w(n) := \log_{10}\left(n^{n+1/2}e^{-n}\right)$ is uniformly distributed mod 1. For every real number $x \ge 1$ define $w(x) = \log_{10}\left(x^{x+1/2}e^{-x}\right)$. Observe that for $1 \le x \le N$ we have

$$w'(x) = \frac{1}{\log(10)}\left(\log x + \frac{1}{2x}\right) \le c\,\log N\,,$$

$$w''(x) = \frac{1}{x \log(10)}\left(1 - \frac{1}{2x}\right) \geq \frac{c}{N} .$$

By applying Lemma 9.9 below we obtain

$$\frac{1}{N}\left|\sum_{n=1}^{N} e^{2\pi i a_n}\right| \leq c \, \frac{\log N}{\sqrt{N}} \longrightarrow 0$$

as $N \to +\infty$. Then, by the Weyl criterion, $w(n)$ is uniformly distributed mod 1 and therefore $n!$ is a strong Benford sequence. $\square$

Remark 6.34 The argument in Remark 6.20 shows that for every non-zero real number b the sequence $b \log_{10}(n)$ is not uniformly distributed, and therefore n^b is not a strong Benford sequence. In particular, the sequence n of the positive integers is not a strong Benford sequence. Of course we cannot deduce that n is not a Benford sequence. We then prove this fact directly. For every positive integer N and every $k \in \{1, 2, \ldots, 9\}$, let

$$q_k(N) = \frac{\text{card}\{n \leq N : k \text{ is the first significant digit of } n\}}{N} .$$

Then a direct computation shows that

$$q_1(9) = \frac{1}{9}, \quad q_1(99) = \frac{11}{99}, \quad q_1(999) = \frac{111}{999}, \quad \ldots$$

so that

$$q_1(10^n - 1) = \frac{1}{9} \neq \log_{10}(2) .$$

Then n is not a Benford sequence. Observe that we have another proof of Remark 6.20.

The study of arithmetic functions in Chapter 2 may suggest replacing $q_1(N)$ with its arithmetic mean $p_1(N) = N^{-1} \sum_{j=1}^{N} q_1(j)$. Unfortunately this idea does not work, since it was proved in [80] that $p_1(N) \nrightarrow \log_{10}(2)$. We may try again and average $p_1(N)$, but we still do not obtain the limit $\log_{10}(2)$. Flehinger [80] has shown that we obtain the correct limits only if we average $q_k(N)$ infinitely many times.

If t_n is a strong Benford sequence then, for every $\alpha > 0$, the same is true for αt_n, so that the law does not depend on the units of measurement (see [139]).

Benford's law can be used to optimize computer data storage, to check demographic projections, to discover scientific and fiscal frauds [37, 122, 132, 133]. It is common to check not only the first digit, but also the first few digits, in a sort of compromise between Benford's law and the strong Benford's law.

Exercises

1) Prove the existence of dense sequences in $\mathbb{T}$ which are not uniformly distributed.

2) Let $t(j)$ be a sequence with values in $\mathbb{T}^d$. Prove that

$$\lim_{N\to\infty}\left(\frac{1}{N}\sum_{j=1}^{N} f(t(j))\right) = \int_{\mathbb{T}^d} f(x)\,dx$$

holds true for every Riemann integrable function f on $\mathbb{T}^d$ if and only if it holds true for every continuous function on $\mathbb{T}^d$.

3) Prove that the sequence $\left(\sqrt{2}j, \left(\sqrt{2}-1\right)j\right)$ is not uniformly distributed in $\mathbb{T}^2$.

4) Prove that we cannot replace the interval I in (6.1) with a Lebesgue measurable set.

5) Prove that

$$\|f\|_{L^\infty(\mathbb{T}^d)} = \lim_{p\to+\infty}\|f\|_{L^p(\mathbb{T}^d)}\ .$$

6) Let $P(t) = \sum a_n e^{2\pi i n\cdot t}$ be a trigonometric polynomial on $\mathbb{T}^d$. Prove that $\widehat{P}(n) = a_n$ for every $n \in \mathbb{Z}^d$.

7) Let $P_N = \sum_{k=0}^{N} a_k\, e^{2\pi i n k x}$ be a trigonometric polynomial on $\mathbb{T}$. Use Proposition 6.10 to prove that

$$\left\|P_N'\right\|_{L^p(\mathbb{T})} \le 2\pi N\, \|P_N\|_{L^p(\mathbb{T})}\ ,$$

where $1 \le p \le \infty$.

8) Let D_N be the Dirichlet kernel (see 6.11). Prove the existence of two positive constants c and c' such that

$$c\,\log N \le \|D_N\|_{L^p(\mathbb{T})} \le c'\,\log N\ .$$

9) Use the argument in the proof of Theorem 4.19 to exhibit a function $f \in C\left(\mathbb{T}^2\right)$ such that the 'summation by squares'

$$S_N^Q f(t) := \sum_{|n_1|\le N,\ |n_2|\le N} \widehat{f}(n)\, e^{2\pi i n\cdot t}$$

converges uniformly as $N \to +\infty$, while

$$\limsup_{N\to+\infty}\left|S_N^D f(0)\right| = +\infty\ ,$$

where

$$S_N^D f(x) := \sum_{|n|\le N} \widehat{f}(n)\, e^{2\pi i n\cdot t}$$

denotes the 'summation by discs'.

10) Let

$$H_M = K_N * K_N * \ldots * K_N$$

be the convolution of M Fejér kernels. Prove that $\|H_M\|_{L^p(\mathbb{T}^d)} = 1$ for every M.

11) Find a Benford sequence which is not a strong Benford sequence.

12) Prove that the Fibonacci sequence is a strong Benford sequence.

13) Let $q_1(N)$ as in Remark 6.34. Compute

$$\liminf_{N \to +\infty} q_1(N) \qquad \text{and} \qquad \limsup_{N \to +\infty} q_1(N) .$$

7

Discrepancy and trigonometric approximation

The definition of uniform distribution (Definition 6.1) does not consider the speed of convergence. In fact an estimate of this speed can be useful, for example, since Theorem 6.4 relates uniformly distributed sequences to the computation of integrals. In this chapter we introduce the definition of discrepancy, which is a quantitative counterpart of the uniform distribution. Then we shall see how to estimate the discrepancy and how to use it to evaluate the approximation of certain integrals by Riemann sums.

Definition 7.1 Let $A_N = \{\omega(j)\}_{j=1}^{N} \subset \mathbb{T}$ be a set of N points. The *discrepancy* of A_N (with respect to intervals) is defined by

$$D_N = D(A_N) = D(\{\omega(j)\}_{j=1}^{N}) := \sup_I \left| |I| \, N - \mathrm{card}\left(\{\omega(j)\}_{j=1}^{N} \cap I\right) \right| ,$$

where the supremum is over all intervals $I \subseteq \mathbb{T}$.

It is sometimes useful to write

$$\frac{D_N}{N} = \sup_I \left| |I| - \frac{1}{N} \mathrm{card}\left(\{\omega(j)\}_{j=1}^{N} \cap I\right) \right|$$

$$= \sup_I \left| \int_{\mathbb{T}} \chi_I(x)\, dx - \frac{1}{N} \sum_{j=1}^{N} \chi_I(\omega(j)) \right|$$

as a sup of distances between integrals and Riemann sums.

It may be difficult to estimate D_N directly, and it is natural to look for bounds related to the exponential sums introduced in the Weyl criterion. The Erdős–Turán inequality is the most important result in this direction (see Section 7.2). We first need a Fourier analytic result on the approximation of the characteristic function of an interval by trigonometric polynomials.

7.1 One-sided trigonometric approximation

The proof of the Weyl criterion uses the completeness of the trigonometric system to approximate the characteristic function χ_I of an interval I with trigonometric polynomials. Here we shall obtain polynomial approximations of χ_I from above and from below, which in turn will lead us to the Erdős–Turán inequality. The proof of the following theorem follows the approach in [58], where a general multidimensional result is proved. We point out the existence of a deeper and sharper version of the following theorem, due to Beurling, Selberg and Vaaler [84, 123, 176].

Theorem 7.2 *There exists a constant $c > 0$ such that for every interval $I \subset \mathbb{T}$ and for every positive integer N there exist two trigonometric polynomials $p_{I,N}^+$ and $p_{I,N}^-$ of degree at most N satisfying*

$$p_{I,N}^-(x) \le \chi_I(x) \le p_{I,N}^+(x)$$

for every $x \in \mathbb{T}$, and

$$\int_{\mathbb{T}} \left(p_{I,N}^+(x) - p_{I,N}^-(x) \right) dx \le \frac{c}{N}. \tag{7.1}$$

Proof It is enough to construct $p_{I,N}^+$ (then we set $p_{I,N}^-(x) := 1 - p_{I^c,N}^+(x)$). Since a translation does not change the degree of a polynomial, we may assume that I is symmetric around the origin and write $I_s := [-s, s]$ (with $0 \le s < 1/2$). We also write $p_{s,N}^+$ in place of $p_{I_s,N}^+$. If $s \ge \frac{1}{2} - \frac{2}{N}$ we can choose $p_{s,N}^+(x) = 1$ for every $x \in \mathbb{T}$. Then we assume that $0 \le s < \frac{1}{2} - \frac{2}{N}$. For every $x \in [-1/2, 1/2)$ we define

$$f_{s,N}(x) := \left(1 + \frac{G}{N^3 (s - |x| + 2/N)^3}\right) \chi_{I_{s+1/N}}(x), \tag{7.2}$$

where the constant $G \ge 1$ will be chosen later during the proof. Note that $f_{s,N}$ is a non-negative function. Let K_M be the Fejér kernel of degree M. By Parseval's identity we have

$$\|K_M\|_{L^2(\mathbb{T})}^2 = \sum_{j=-M}^{M} \left(1 - \frac{|j|}{M+1}\right)^2 = 1 + \frac{2}{(M+1)^2} \sum_{j=1}^{M} (M + 1 - j)^2$$

$$= 1 + \frac{2}{(M+1)^2} \sum_{j=1}^{M} j^2 = 1 + \frac{M(2M+1)}{3(M+1)}.$$

Hence $\|K_M\|_{L^2(\mathbb{T})} \approx \sqrt{M}$, where $A \approx B$ means that $A > 0$, $B > 0$ and there exist two positive constants c_1 and c_2 such that $c_1 A \le B \le c_2 A$. We may assume that

N is even, say $N = 2M$, with $M \geq 2$. We introduce the *Jackson kernel*

$$\mathcal{J}_N(x) := \frac{1}{\|K_M\|_{L^2(\mathbb{T})}^2} K_M^2(x) = \frac{1}{\|K_M\|_{L^2(\mathbb{T})}^2} \frac{\sin^4(\pi(M+1)x)}{(M+1)^2 \sin^4(\pi x)} \qquad (7.3)$$

$$\approx \frac{\sin^4(\pi(M+1)x)}{N^3 \sin^4(\pi x)} \leq c \min\left(N, \frac{1}{N^3 |x|^4}\right).$$

The Jackson kernel resembles the Fejér kernel, but it is more concentrated near the origin, see [105] and Lemma 6.11. Since $\widehat{K}_M(m) \geq 0$ for every integer m, we have

$$\|K_M\|_{L^2(\mathbb{T})}^2 \, \widehat{\mathcal{J}}_N(k) = \int_{\mathbb{T}} K_M^2(x) \, e^{-2\pi i k x} \, dx$$

$$= \int_{\mathbb{T}} \left(\sum_{m=-M}^{M} \widehat{K}_M(m) \, e^{2\pi i m x}\right)^2 e^{-2\pi i k x} \, dx$$

$$= \int_{\mathbb{T}} \sum_{m,n=-M}^{M} \widehat{K}_M(m) \widehat{K}_M(n) \, e^{2\pi i (m+n-k)x} \, dx$$

$$= \sum_{m,n=-M}^{M} \widehat{K}_M(m) \widehat{K}_M(n) \int_{\mathbb{T}} e^{2\pi i (m+n-k)x} \, dx$$

$$= \sum_{m=-M}^{M} \widehat{K}_M(m) \widehat{K}_M(k-m) \geq 0.$$

Therefore for every $k \in \mathbb{Z}$ we have

$$0 \leq \widehat{\mathcal{J}}_N(k) \leq \|\mathcal{J}_N\|_{L^1(\mathbb{T})} = \int_{\mathbb{T}} \mathcal{J}_N = \frac{1}{\|K_M\|_{L^2(\mathbb{T})}^2} \int_{\mathbb{T}} K_M^2 = 1. \qquad (7.4)$$

The trigonometric polynomial (see (7.2))

$$p_{I_s,N}^+ := p_{s,N}^+ := f_{s,N} * \mathcal{J}_N$$

has degree $\leq N$. Let us show that $p_{s,N}^+(x) \geq \chi_{I_s}(x)$ for every $x \in \mathbb{T}$. Since $p_{s,N}^+$ is non-negative and symmetric around the origin, we only have to prove that $p_{s,N}^+(x) \geq 1$ for every $0 \leq x \leq s$. By (7.4) this means proving that we have

$$\int_{\mathbb{T}} \mathcal{J}_N(y) f_{s,N}(x-y) \, dy \geq \int_{\mathbb{T}} \mathcal{J}_N(y) \, dy = 1$$

for every $0 \leq x \leq s$, that is,

$$\int_{\mathbb{T}} \mathcal{J}_N(y) (f_{s,N}(x-y) - 1) \, dy \geq 0,$$

so

$$\int_{x+I_{s+1/N}} \mathcal{J}_N(y) \frac{G}{N^3 (s - |x - y| + 2/N)^3} \, dy \geq \int_{(x+I_{s+1/N})^c} \mathcal{J}_N(y) \, dy . \quad (7.5)$$

We are going to prove (7.5). In the rest of this proof $c, c_1, c_2, \ldots$ will be positive constants independent of s, N, G. Let us call D_x and E_x, respectively, the LHS and the RHS of (7.5). In order to prove that $D_x \geq E_x$ we first observe that if $|y| \leq \frac{1}{N}$ and $0 \leq x \leq s$, then $y \in (x + I_{s+1/N})$. Therefore (7.3) implies

$$D_x \geq \int_{-1/N}^{1/N} \mathcal{J}_N(y) \frac{G}{N^3 (s - |x - y| + 2/N)^3} \, dy$$

$$\geq \min_{|y| \leq 1/N} \left(\frac{G}{N^3 (s - |x - y| + 2/N)^3} \right) \frac{1}{\|K_M\|_{L^2(\mathbb{T})}^2} \int_{-1/N}^{1/N} K_M^2(y) \, dy$$

$$\geq c_1 \frac{G}{N^6 (s - x + 3/N)^3} \int_0^{1/N} \frac{\sin^4 (\pi (M + 1) y)}{\sin^4 (\pi y)} \, dy$$

$$\geq c_2 \frac{G}{N^3 (s - x + 3/N)^3} ,$$

since $\sin (\pi (M + 1) y) \geq c_3 N y$ when $0 \leq y \leq 1/N$. Now observe that if $|y| \leq s - x + 1/N$, then $y \in (x + I_{s+1/N})$. Then (7.3) implies

$$E_x \leq \int_{|y| > s-x+1/N} \mathcal{J}_N(y) \, dy \leq c_4 \frac{1}{N^3} \int_{s-x+1/N}^{1/2} \frac{1}{y^4} \, dy \leq c_5 \frac{1}{N^3 \left(s + \frac{1}{N} - x\right)^3} .$$

If we choose $G = 27c_5/c_2$, then $E_x \leq D_x$ for every $0 \leq x \leq s$. We now prove (7.1). It is enough to estimate

$$\int_{\mathbb{T}} \left(p_{s,N}^+ - \chi_{I_s} \right) = \int_{\mathbb{T}} p_{s,N}^+ - 2s .$$

Since $p_{s,N}$ is a non-negative function, we can apply Proposition 6.10 and (7.4) to obtain

$$\int_{\mathbb{T}} p_{s,N}^+ = \left\| f_{s,N} * \mathcal{J}_N \right\|_{L^1(\mathbb{T})} \leq \left\| f_{s,N} \right\|_{L^1(\mathbb{T})}$$

$$= \int_{\mathbb{T}} \left(1 + \frac{G}{N^3 (s - |x| + 2/N)^3} \right) \chi_{I_{s+1/N}}(x) \, dx$$

$$= 2s + 2/N + 2G \int_0^{s+1/N} \frac{1}{N^3 (s - x + 2/N)^3} \, dx .$$

Hence

$$\int_{\mathbb{T}} \left(p_{I,N}^+ - \chi_I \right) \leq \frac{2}{N} + 2G \frac{1}{N} \int_1^{Ns+2} \frac{1}{u^3} \, du \leq \frac{c_6}{N} .$$

$\square$

7.2 The Erdős–Turán inequality

We can now prove the following result (see [77]).

Theorem 7.3 (Erdős–Turán inequality) *There exists $c > 0$ such that for every choice of positive integers N, H and for every finite sequence $\omega = \{\omega(j)\}_{j=1}^{N} \subset \mathbb{T}$ we have*

$$D_N \le c \left(\frac{N}{H} + \sum_{k=1}^{H} \frac{1}{k} \left| \sum_{j=1}^{N} e^{2\pi i k \omega(j)} \right| \right). \tag{7.6}$$

Proof Let I be an interval in $\mathbb{T}$. By Theorem 7.2 there exists a trigonometric polynomial $p_{I,H}^{+}$ having degree at most H and satisfying

$$\text{card}\left(\{\omega(j)\}_{j=1}^{N} \cap I \right)$$
$$= \sum_{j=1}^{N} \chi_I(\omega(j)) \le \sum_{j=1}^{N} p_{I,H}^{+}(\omega(j)) = \sum_{j=1}^{N} \sum_{k=-H}^{H} \widehat{p_{I,H}^{+}}(k)\, e^{2\pi i k \omega(j)}$$
$$= \sum_{k=-H}^{H} \widehat{p_{I,H}^{+}}(k) \sum_{j=1}^{N} e^{2\pi i k \omega(j)} \le \left(|I| + \frac{c}{H} \right) N + \sum_{0 < |k| \le H} \widehat{p_{I,H}^{+}}(k) \sum_{j=1}^{N} e^{2\pi i k \omega(j)}$$

(a similar bound holds for $p_{I,H}^{-}$). For $0 < |k| \le H$ we have

$$\left| \widehat{p_{I,H}^{\pm}}(k) \right| \le \left| \widehat{p_{I,H}^{\pm}}(k) - \widehat{\chi_I}(k) \right| + \left| \widehat{\chi_I}(k) \right| \le \left\| p_{I,H}^{\pm} - \chi_I \right\|_{L^1(\mathbb{T})} + \frac{1}{\pi |k|}$$
$$\le \frac{c_2}{H} + \frac{1}{\pi |k|} \le \frac{c_3}{|k|},$$

since it is easy to see that for every interval $I \subseteq \mathbb{T}$ and every integer $k \ne 0$ we have

$$\left| \widehat{\chi_I}(k) \right| \le \frac{1}{\pi |k|}. \tag{7.7}$$

Then the proof of (7.6) is complete. $\square$

We now apply the Erdős–Turán inequality to estimate the discrepancy of certain finite sequences $\{j\alpha\}_{j=1}^{N} \subset \mathbb{T}$.

For every $x \in \mathbb{R}$ let

$$\|x\| := \min\left(\{x\}, 1 - \{x\} \right)$$

be the distance from x to the integers.

Theorem 7.4 *Let α be an irrational algebraic number of degree 2. Then there* exists a positive constant c such that for every integer $N \geq 2$ and $\omega = \omega(j) = \{j\alpha\}_{j=1}^{N}$ we have

$$D_N = D_N(\omega) \leq c \log^2 N . \tag{7.8}$$

Proof Let us choose $H = N$ in the Erdős–Turán inequality (7.6). Since

$$|\sin(\pi x)| = \sin(\pi \|x\|) \geq 2 \|x\|$$

we obtain

$$D_N(\omega) \leq c\left(1 + \sum_{k=1}^{N} \frac{1}{k} \left|\sum_{j=1}^{N} e^{2\pi i k j\alpha}\right|\right) \leq c\left(1 + \sum_{k=1}^{N} \frac{1}{k} \frac{2}{\left|e^{2\pi i k \alpha} - 1\right|}\right)$$

$$= c\left(1 + \sum_{k=1}^{N} \frac{1}{k \,|\sin(\pi k \alpha)|}\right) \leq c\left(1 + \sum_{k=1}^{N} \frac{1}{k \,\|k\alpha\|}\right) .$$

By the *partial summation formula*[1] we obtain

$$D_N(\omega) \leq c\left(1 + \sum_{k=1}^{N-1} \frac{1}{k(k+1)} \sum_{j=1}^{k} \frac{1}{\|j\alpha\|} + \frac{1}{N} \sum_{j=1}^{N} \frac{1}{\|j\alpha\|}\right) . \tag{7.10}$$

Now we shall prove the existence of a positive constant c_α such that, for every integer $L \geq 2$,

$$\sum_{j=1}^{L} \frac{1}{\|j\alpha\|} \leq c_\alpha \, L \log L . \tag{7.11}$$

Indeed, Theorem 5.6 implies the existence of a positive constant $H = H(\alpha)$ such that

$$\|j\alpha\| \geq \frac{H}{j} \tag{7.12}$$

for every j. We claim that, for every integer $1 \leq s \leq L/(2H)$ and for every $1 \leq j \leq L$, the interval $[sH/L, (s+1)H/L)$ contains at most two values of $\|j\alpha\|$. Assume that there are three of them. Then we have, say,

$$\frac{sH}{L} \leq \{j_1\alpha\}, \{j_2\alpha\} < \frac{(s+1)H}{L} .$$

---

[1] If $A_k = \sum_{j=1}^{k} a_j$, then

$$\sum_{k=1}^{N} a_k b_k = A_1 b_1 + \sum_{k=2}^{N} (A_k - A_{k-1}) b_k = A_1 b_1 + \sum_{k=2}^{N} A_k b_k - \sum_{k=1}^{N-1} A_k b_{k+1} \tag{7.9}$$

$$= A_1 b_1 + A_N b_N + \sum_{k=2}^{N-1} A_k (b_k - b_{k+1}) - A_1 b_2 = A_N b_N + \sum_{k=1}^{N-1} A_k (b_k - b_{k+1}) .$$

We assume that $j_1 < j_2$. Since $|\{j_2\alpha\} - \{j_1\alpha\}| \leq 1/2$ we have

$$\|(j_2 - j_1)\alpha\| = |\{j_2\alpha\} - \{j_1\alpha\}| < \frac{H}{L} \leq \frac{H}{j_2 - j_1} \, ,$$

contrary to (7.12). In the same way we see that the interval $[0, H/L)$ is empty. Then, assuming $L \geq 4H$,

$$\sum_{j=1}^{L} \frac{1}{\|j\alpha\|} \leq 2 \sum_{s=1}^{L/(2H)} \frac{L}{sH} \leq c_\alpha \, L \log L \, .$$

Since there is a bounded number of integers $L < 4H$, then (7.11) is proved for every L. Hence (7.10) and (7.11) imply

$$D_N(\omega) \leq c_2 \left(\log N + \sum_{k=1}^{N-1} \frac{\log k}{(k+1)} \right) \leq c_3 \log^2 N \, .$$

$\square$

The inequality in (7.8) can be improved. See [114, p. 125], where $\log^2 N$ is replaced with $\log N$.

7.3 The inequalities of Koksma and Koksma–Hlawka

It is possible to use upper bounds of discrepancies to estimate approximations of integrals by Riemann sums. In one variable, the most important result of this type is due to Koksma [109]. We first recall the definition of bounded variation.

Definition 7.5 A function $f : [0, 1] \to \mathbb{R}$ is said to be of *bounded variation* on $[0, 1]$ if

$$V_f := \sup_{\substack{0 \leq x_0 \leq x_1 \leq x_2 \leq \ldots \leq x_P \leq 1, \\ P \in \mathbb{N}}} \left(\sum_{\ell=1}^{P} |f(x_\ell) - f(x_{\ell-1})| \right) < +\infty \, .$$

We also recall that for a smooth function f on $[0, 1]$ we have

$$V_f = \int_0^1 |f'(x)| \, dx \, ,$$

see [182].

Lemma 7.6 (Jordan's theorem) *Let f be a function of bounded variation on $[0, 1]$. Then there exist two increasing functions g, h such that $f = g - h$ and $V_f = V_g + V_h$.*

Proof The function

$$V_f(x) := \sup_{\substack{0 \le x_0 \le x_1 \le x_2 \le \ldots \le x_{P-1} \le x_P = x, \\ P \in \mathbb{N}}} \left(\sum_{\ell=1}^{P} |f(x_\ell - f(x_{\ell-1}))| \right)$$

is increasing on $[0, 1]$ and satisfies $V_f(0) = 0$ and $V_f(1) = V_f$. Let $\varepsilon > 0$, choose P and $\overline{x}_0, \overline{x}_1, \overline{x}_2, \ldots, \overline{x}_{P-1}, \overline{x}_P = x$ such that

$$\sum_{\ell=1}^{P} |f(\overline{x}_\ell) - f(\overline{x}_{\ell-1})| \ge V_f(x) - \varepsilon .$$

Let $x < y \le 1$. Then

$$V_f(y) \pm f(y)$$

$$\ge \left(|f(y) - f(x)| + \sum_{\ell=1}^{P} |f(\overline{x}_\ell - f(\overline{x}_{\ell-1}))| \right) \pm (f(y) - f(x) + f(x))$$

$$\ge V_f(x) - \varepsilon \pm f(x) .$$

Hence $V_f(y) \pm f(y) \ge V_f(x) \pm f(x)$. Then the functions $t \to V_f(t) \pm f(t)$ are increasing. If we write

$$f(t) = \frac{V_f(t) + f(t)}{2} - \frac{V_f(t) - f(t)}{2} ,$$

we prove the first part of the lemma. Finally observe that an increasing function $g : [0, 1] \to \mathbb{R}$ satisfies $V_g = g(1) - g(0)$. Then

$$V_{(V_f+f)/2} + V_{(V_f-f)/2}$$

$$= \frac{1}{2}(V_f + f)(1) - \frac{1}{2}(V_f + f)(0) + \frac{1}{2}(V_f - f)(1) - \frac{1}{2}(V_f - f)(0)$$

$$= V_f(1) - V_f(0) = V_f .$$

$\square$

The following definition is essentially equivalent to Definition 7.1, since it is easy to prove that $D_N^* \le D_N \le 2D_N^*$.

Definition 7.7 Let

$$D_N^* = D_N^* \left(\{\omega(j)\}_{j=1}^{N} \right) := \sup_{\alpha \in [0,1]} \left| \mathrm{card} \left([0, \alpha] \cap \{\omega(j)\}_{j=1}^{N} \right) - N\alpha \right|$$

$$= \sup_{\alpha \in [0,1]} \left| \sum_{j=1}^{N} \chi_{[0,\alpha]}(\omega(j)) - N \int_0^1 \chi_{[0,\alpha]}(x) \, dx \right| .$$

Theorem 7.8 (Koksma's inequality)   *Let $\{\omega(j)\}_{j=1}^{N} \subset [0, 1]$. Then for every function f of bounded variation on $[0, 1]$ we have*

$$\left| \frac{1}{N} \sum_{j=1}^{N} f(\omega(j)) - \int_{0}^{1} f(x)\, dx \right| \leq V_f \, \frac{D_N^*}{N} .$$

Proof   The previous lemma allows us to assume that f is a decreasing function satisfying $f(0) = 1$ and $f(1) = 0$. Choose a large positive integer M, let

$$I_1 = \left\{ x \in [0, 1] : 1 - \frac{1}{M} \leq f(x) \leq 1 \right\}$$

and, for each $h = 2, \ldots, M$, consider the interval

$$I_h = \left\{ x \in [0, 1] : \frac{M - h}{M} \leq f(x) < \frac{M - h + 1}{M} \right\} .$$

Then we have the disjoint union $[0, 1] = \cup_{h=1}^{M} I_h$. Note that the intervals I_h possibly reduce to one point or to the empty set. Let

$$S_M(x) := \sum_{h=1}^{M} \frac{M - h + 1}{M} \chi_{I_h}(x) = \frac{1}{M} \sum_{h=1}^{M} \chi_{\bar{I}_h}(x) , \qquad (7.13)$$

where every $\bar{I}_h := \cup_{s=1}^{h} I_s$ is an interval anchored at 0. Then

$$\left| \frac{1}{N} \sum_{j=1}^{N} f(\omega(j)) - \int_{0}^{1} f(x)\, dx \right| \qquad (7.14)$$

$$\leq \left| \frac{1}{N} \sum_{j=1}^{N} S_M(\omega(j)) - \int_{0}^{1} S_M(x)\, dx \right|$$

$$+ \frac{1}{N} \left| \sum_{j=1}^{N} (f(\omega(j)) - S_M(\omega(j))) \right| + \left| \int_{0}^{1} (f(x) - S_M(x))\, dx \right| .$$

By (7.13) we have

$$\left| \frac{1}{N} \sum_{j=1}^{N} S_M(\omega(j)) - \int_{0}^{1} S_M(x)\, dx \right|$$

$$\leq \frac{1}{M} \sum_{h=1}^{M} \left| \frac{1}{N} \sum_{j=1}^{N} \chi_{\bar{I}_h}(\omega(j)) - \int_{0}^{1} \chi_{\bar{I}_h}(x)\, dx \right| \leq \frac{1}{M} \sum_{h=1}^{M} \frac{D_N^*}{N} = V_f \frac{D_N^*}{N}$$

(since $V_f = 1$). Now we consider the second term in the RHS of (7.14). Since

$$\chi_{[0,1]} = \sum_{h=1}^{M} \chi_{I_h}$$

we have

$$\frac{1}{N}\left|\sum_{j=1}^{N}(f(\omega(j)) - S_M(\omega(j)))\right| = \frac{1}{N}\left|\sum_{j=1}^{N}\left(f(\omega(j)) - \sum_{h=1}^{M}\frac{M-h+1}{M}\chi_{I_h}(\omega(j))\right)\right|$$

$$= \frac{1}{N}\sum_{j=1}^{N}\left(\sum_{h=1}^{M}\chi_{I_h}(\omega(j))\left(\frac{M-h+1}{M} - f(\omega(j))\right)\right)$$

$$\leq \frac{1}{M}\frac{1}{N}\sum_{j=1}^{N}\left(\sum_{h=1}^{M}\chi_{I_h}(\omega(j))\right) = \frac{1}{M}.$$

Finally

$$\left|\int_{0}^{1}(f(x) - S_M(x))\,dx\right| = \left|\int_{0}^{1}\left(f(x) - \sum_{h=1}^{M}\frac{M-h+1}{M}\chi_{I_h}(x)\right)dx\right|$$

$$= \sum_{h=1}^{M}\int_{I_h}\left(\frac{M-h+1}{M} - f(x)\right)dx \leq \frac{1}{M}.$$

This completes the proof since M is an arbitrarily large number.   □

By Theorem 7.4 we readily obtain the following result.

Corollary 7.9   *Let α be an irrational algebraic number of degree 2 and let f be a function of bounded variation on $[0,1]$. Then there exists a positive constant c such that*

$$\left|\frac{1}{N}\sum_{j=1}^{N}f(\{j\alpha\}) - \int_{T}f(t)dt\right| \leq c\,V_f\,\frac{\log^2 N}{N}.$$

We may say that Koksma's inequality turns the discrepancy for a small class of functions (characteristic functions of intervals) into the discrepancy for a large class (functions of bounded variation). See [22, 24, 25, 26, 91, 114]. We wish to show that it may be natural to choose (a mild modification of) the sawtooth function $s(x)$ (see (4.33)) and its translates as the 'small class of functions'.

We consider the periodic function

$$g(x) := \frac{1}{2} - s(x) \tag{7.15}$$

(then $g(x) = 1 - x$ if $0 \le x < 1$). For $x \notin \mathbb{Z}$ we have (see (4.34))

$$g(x) = \int_0^1 \chi_{[0,\alpha]+\mathbb{Z}}(x)\, d\alpha = \frac{1}{2} + \sum_{k \ne 0} \frac{1}{2\pi ik} e^{2\pi ikx} \,.$$

Remark 7.10   The function $g(x)$ has a peculiar property. On the one hand, the convolution with $g(x)$ is the inverse of the differentiation. Indeed, for $k \ne 0$, integration by parts yields

$$\widehat{g}(k)\, \widehat{f'}(k) = \frac{1}{2\pi ik} \int_{\mathbb{T}} f'(x) e^{-2\pi ikx}\, dx = \int_{\mathbb{T}} f(x)\, e^{-2\pi ikx}\, dx = \widehat{f}(k) \,.$$

On the other hand, $g(x)$ is a superposition of characteristic functions of intervals anchored at the origin.

Theorem 7.11   *Let $\{t(j)\}_{j=1}^N \subset \mathbb{T}$ and let $g(x)$ be as in (7.15). Let f be a smooth function on $\mathbb{T}$. Then*

$$\sup_{x \in \mathbb{T}} \left| \frac{1}{N} \sum_{j=1}^N f(x + t(j)) - \int_{\mathbb{T}} f(y)\, dy \right| \tag{7.16}$$

$$\le \left(\inf_{a \in \mathbb{R}} \int_{\mathbb{T}} |a + f'(x)|\, dx \right) \left(\sup_{x \in \mathbb{T}} \left| \frac{1}{N} \sum_{j=1}^N g(x + t(j)) \right| \right) .$$

Proof   The periodicity of f and integration by parts yield

$$\widehat{f}(k) = \int_0^1 f(x) e^{-2\pi ikx}\, dx = \frac{1}{2\pi ik} \int_0^1 f'(x) e^{-2\pi ikx}\, dx = \frac{1}{2\pi ik} \widehat{f'}(k)$$

$$= \frac{1}{-4\pi^2 k^2} \widehat{f''}(k) = \frac{1}{-8\pi^3 ik^3} \widehat{f'''}(k)$$

for every $k \ne 0$. Hence the Fourier series of $f(x)$ and $f'(x)$ converge absolutely, then they converge pointwise to $f(x)$ and $f'(x)$, respectively. Observe that the function $x \mapsto f(x + t(j))$ has Fourier coefficients

$$f(\cdot + t(j))^\wedge(k) = \int_{\mathbb{T}} f(x + t(j))\, e^{-2\pi ikx}\, dx = e^{2\pi ikt(j)} \int_{\mathbb{T}} f(u)\, e^{-2\pi iku}\, du$$

$$= e^{2\pi ikt(j)} \widehat{f}(k)$$

(in particular $f(\cdot + t(j))^\wedge(0) = \widehat{f}(0)$). By (7.15) we have, for every $a \in \mathbb{R}$,

$$\sup_{x \in \mathbb{T}} \left| \frac{1}{N} \sum_{j=1}^N f(x + t(j)) - \int_{\mathbb{T}} f(y) dy \right| = \sup_{x \in \mathbb{T}} \left| \sum_{k \ne 0} \left(\frac{1}{N} \sum_{j=1}^N e^{2\pi ikt(j)} \right) \widehat{f}(k)\, e^{2\pi ikx} \right|$$

$$= \sup_{x \in \mathbb{T}} \left| \sum_{k \ne 0} \left(\frac{1}{2\pi ik} \frac{1}{N} \sum_{j=1}^N e^{2\pi ikt(j)} \right) (2\pi ik)\, \widehat{f}(k)\, e^{2\pi ikx} \right|$$

$$= \sup_{x\in\mathbb{T}} \left| \sum_{k\neq 0} \left(\widehat{g}(k) \frac{1}{N} \sum_{j=1}^{N} e^{2\pi i k t(j)} \right) (a+f')^\wedge (k)\, e^{2\pi i k x} \right|$$

$$= \sup_{x\in\mathbb{T}} \left| \left(\left(\frac{1}{N} \sum_{j=1}^{N} g\left(\cdot + t(j) \right) \right) * (a + f') \right)(x) \right| .$$

Hence (7.16) follows from Proposition 6.10.   □

Koksma's inequality deals with functions of bounded variation and actually most (bounded) familiar functions on $\mathbb{T}$ share this property. Now we want to introduce a multidimensional extension of Koksma's inequality (results of this type are usually termed *Koksma–Hlawka inequalities*, see e.g. [94, 95, 114, 121, 131]) which holds for a reasonably large class of functions (possibly non-continuous).

Definition 7.12 We say that a function $h(t)$ on $\mathbb{T}^d$ is *piecewise smooth*[2] on $\mathbb{T}^d$ $(d > 1)$ if we can write $h(t) = f(t)\chi_\Omega(t)$, where f is smooth (i.e., it has derivatives of all orders) on $\mathbb{T}^d$ and $\chi_\Omega(t)$ is the characteristic function of an open set Ω in $\mathbb{T}^d$.

Lemma 7.13 Let $f \in C^1\left(\mathbb{T}^d\right)$. Then for $x = (x_1,\ldots,x_d)$ and $n = (n_1,\ldots,n_d)$ we have

$$\left(\frac{\partial f}{\partial x_k} \right)^\wedge (n) = 2\pi i n_k\, \widehat{f}(n)$$

for every $k = 1,\ldots,d$.

Proof We may assume that $k = 1$. Integration by parts gives

$$\int_{\mathbb{T}^d} \frac{\partial f}{\partial x_1}(x)\, e^{-2\pi i n\cdot x}\, dx$$

$$= \int_0^1 \cdots \int_0^1 \left(-\int_0^1 f(x)(-2\pi i n_1)\, e^{-2\pi i n_1 x_1}\, dx_1 \right) e^{-2\pi i (n_2 x_2 + \ldots + n_d x_d)}\, dx_2 \cdots dx_d$$

$$= 2\pi i n_1 \int_{\mathbb{T}^d} f(x)\, e^{-2\pi i n\cdot x}\, dx = 2\pi i n_1\, \widehat{f}(n) .$$

□

The following result was proved in [25]. For simplicity we state and prove it only in the planar case.

---

[2] Note that when $d = 1$ this definition does not coincide with Definition 4.17.

Theorem 7.14 *Let* $\{t(j)\}_{j=1}^{N} \subset \mathbb{T}^2$ *and let* $h(t) = f(t)\chi_\Omega(t)$ *be a piecewise smooth function on* $\mathbb{T}^2$. *Let* $1 \le p, q \le +\infty$ *satisfy* $1/p + 1/q = 1$. *Let*

$$\mathcal{V}_q(f) = 4\|f\|_{L^q(\mathbb{T}^d)} + 2\left\|\frac{\partial f}{\partial t_1}\right\|_{L^q(\mathbb{T}^d)} + 2\left\|\frac{\partial f}{\partial t_2}\right\|_{L^q(\mathbb{T}^d)} + \left\|\frac{\partial^2 f}{\partial t_1 \partial t_2}\right\|_{L^q(\mathbb{T}^d)}.$$

Let $I_s = [0, s_1] \times [0, s_2]$ *be an interval anchored at the origin and let*

$$D_p := \int_{[0,1]^2} \left\|\frac{1}{N}\sum_{j=1}^{N} \chi_{(-I_{s+\mathbb{Z}^2})\cap\Omega}(t(j)) - |(\cdot - I_{s+\mathbb{Z}^2})\cap\Omega|\right\|_{L^p(\mathbb{T}^d)} ds,$$

where $|\cdot|$ *denotes the volume. Then*

$$\left|\frac{1}{N}\sum_{j=1}^{N} h(t(j)) - \int_{\mathbb{T}^2} h(v)\,dv\right| \le \mathcal{V}_q(f)\,D_p.$$

Proof Let $g(t)$ be the periodic function

$$g(t) := \int_{[0,1]^2} \chi_{I_{u+\mathbb{Z}^2}}(t)\,du. \tag{7.17}$$

Then

$$\widehat{g}(n) = \int_0^1 \int_0^1 (1-t_1)(1-t_2)\,e^{-2\pi i(n_1 t_1 + n_2 t_2)}\,dt_1 dt_2 \tag{7.18}$$

$$= \frac{1}{2\delta(n_1) + 2\pi i n_1}\,\frac{1}{2\delta(n_2) + 2\pi i n_2},$$

where

$$\delta(n_k) = \begin{cases} 1 & \text{if } n_k = 0, \\ 0 & \text{if } n_k \ne 0. \end{cases}$$

We introduce the operator $\mathfrak{D}$ defined by

$$\mathfrak{D}f(t) := \sum_{n\in\mathbb{Z}^2} ((2\delta(n_1) - 2\pi i n_1)(2\delta(n_2) - 2\pi i n_2))\,\widehat{f}(n)\,e^{2\pi i n \cdot t}. \tag{7.19}$$

We observe that

$$\left\{\int_{\mathbb{T}^2} |\mathfrak{D}f(t)|^q dt\right\}^{1/q} \le \mathcal{V}_q(f). \tag{7.20}$$

Indeed, for every $t = (t_1, t_2) \in \mathbb{T}^2$ Lemma 7.13 implies

$$\mathfrak{D}f(t_1, t_2) = 4\widehat{f}(0,0) - 2\sum_{n_1\in\mathbb{Z}} 2\pi i n_1\,\widehat{f}(n_1, 0)\,e^{2\pi i n_1 t_1}$$

$$- 2\sum_{n_2\in\mathbb{Z}} 2\pi i n_2\,\widehat{f}(0, n_2)\,e^{2\pi i n_2 t_2} + \sum_{n_1, n_2\in\mathbb{Z}} 2\pi i n_1\,2\pi i n_2\,\widehat{f}(n_1, n_2)\,e^{2\pi i(n_1 t_1 + n_2 t_2)}$$

$$
= 4\widehat{f}(0,0) - 2 \sum_{n_1,n_2 \in \mathbb{Z}} 2\pi i n_1 \, \widehat{f}(n_1,n_2) \, e^{2\pi i n_1 t_1} \int_0^1 e^{2\pi i n_2 t_2} \, dt_2
$$

$$
- 2 \sum_{n_1,n_2 \in \mathbb{Z}} 2\pi i n_2 \, \widehat{f}(n_1,n_2) \, e^{2\pi i n_2 t_2} \int_0^1 e^{2\pi i n_1 t_1} \, dt_1 + \frac{\partial^2 f}{\partial t_1 \partial t_2}(t_1,t_2)
$$

$$
= 4 \int_{[0,1]^2} f(t_1,t_2) \, dt_1 dt_2 - 2 \int_0^1 \frac{\partial f}{\partial t_1}(t_1,t_2) \, dt_2
$$

$$
- 2 \int_0^1 \frac{\partial f}{\partial t_2}(t_1,t_2) \, dt_1 + \frac{\partial^2 f}{\partial t_1 \partial t_2}(t_1,t_2) \,,
$$

which implies (7.20). Observe that Theorem 4.15, (7.17), (7.18) and (7.19) imply

$$
\frac{1}{N} \sum_{j=1}^N f(t(j)) \chi_\Omega(t(j)) - \int_{\mathbb{T}^2} f(t) \chi_\Omega(t) \, dt
$$

$$
= \frac{1}{N} \sum_{j=1}^N \left(\sum_{n \in \mathbb{Z}^2} \widehat{f}(n) \, e^{2\pi i n \cdot t(j)} \right) \chi_\Omega(t(j)) - \sum_{n \in \mathbb{Z}^2} \widehat{f}(n) \overline{\widehat{\chi_\Omega}(n)}
$$

$$
= \frac{1}{N} \sum_{j=1}^N \left(\sum_{n \in \mathbb{Z}^2} \widehat{(\mathfrak{D}f)}(n) \, \overline{\widehat{g_{-t(j)}}(n)} \right) \chi_\Omega(t(j)) - \sum_{n \in \mathbb{Z}^2} \widehat{(\mathfrak{D}f)}(n) \overline{\widehat{g}(n) \widehat{\chi_\Omega}(n)}
$$

$$
= \frac{1}{N} \sum_{j=1}^N \int_{\mathbb{T}^2} \mathfrak{D}f(u) \, g_{-t(j)}(u) \, du \, \chi_\Omega(t(j)) - \int_{\mathbb{T}^2} \mathfrak{D}f(u) \, (g * \chi_\Omega)(u) \, du
$$

$$
= \int_{\mathbb{T}^2} \mathfrak{D}f(u) \left(\frac{1}{N} \sum_{j=1}^N \chi_\Omega(t(j)) g(u - t(j)) - \int_{\mathbb{T}^2} \chi_\Omega(t) g(u - t) \, dt \right) du \,,
$$

where $g_v(u) := g(u + v)$. Therefore Hölder's inequality, (7.20), (7.17) and Minkowski's integral inequality imply

$$
\left| \frac{1}{N} \sum_{j=1}^N f(t(j)) \chi_\Omega(t(j)) - \int_{\mathbb{T}^2} f(t) \chi_\Omega(t) \, dt \right|
$$

$$
\leq \left\{ \int_{\mathbb{T}^2} |\mathfrak{D}f(t)|^q \, dt \right\}^{1/q}
$$

$$
\times \left\{ \int_{\mathbb{T}^2} \left| \frac{1}{N} \sum_{j=1}^N \chi_\Omega(t(j)) g(u - t(j)) - \int_{\mathbb{T}^2} \chi_\Omega(t) g(u - t) \, dt \right|^p du \right\}^{1/p}
$$

$$
\leq \mathcal{V}_q(f) \left\{ \int_{\mathbb{T}^2} \left| \frac{1}{N} \sum_{j=1}^N \chi_\Omega(t(j)) \int_{[0,1]^2} \chi_{I_{s+\mathbb{Z}^2}}(u - t(j)) \, ds \right. \right.
$$

$$- \int_{\mathbb{T}^2} \chi_\Omega(t) \int_{[0,1]^2} \chi_{I_{s+\mathbb{Z}^2}}(u-t) \, ds \Big|^p \, du \Big\}^{1/p}$$

$$\leq \mathcal{V}_q(f) \int_{[0,1]^2} \Big\{ \int_{\mathbb{T}^2} \Big| \frac{1}{N} \sum_{j=1}^{N} \chi_\Omega(t(j)) \chi_{I_{s+\mathbb{Z}^2}}(u-t(j))$$

$$- \int_{\mathbb{T}^2} \chi_\Omega(t) \chi_{I_{s+\mathbb{Z}^2}}(u-t) \Big|^p \, du \Big\}^{1/p} ds$$

$$= \mathcal{V}_q(f)$$

$$\times \int_{[0,1]^2} \Big\{ \int_{\mathbb{T}^2} \Big| \frac{1}{N} \sum_{j=1}^{N} \chi_{(u-I_{s+\mathbb{Z}^2}) \cap \Omega}(t(j)) - |(u - I_{s+\mathbb{Z}^2}) \cap \Omega| \Big|^p \, du \Big\}^{1/p} ds$$

$$= \mathcal{V}_q(f) D_p \, .$$

$$\square$$

The previous theorem has introduced L^p norms of multidimensional discrepancies with respect to fairly general sets. This will be one of our main interests in the last chapters.

Exercises

1) Let N be a large positive integer. Prove the existence of an interval $I \subset \mathbb{T}$ such that, for every trigonometric polynomial $T(x)$ of degree N, satisfying $T(x) \geq \chi_I(x)$ for every $x \in \mathbb{T}$, we have

$$\int_{\mathbb{T}} (T - \chi_I) \geq \frac{1}{N+1} \, .$$

2) Prove that an infinite sequence $\{\omega(j)\}_{j=1}^{\infty} \subset \mathbb{T}$ is uniformly distributed if and only if

$$D\left(\{\omega(j)\}_{j=1}^{N}\right) = o(N)$$

as $N \to \infty$.

3) Let $s(x)$ be the sawtooth function (see (4.33)) and let α be an irrational algebraic number of degree 2. Prove the inequality

$$\Big| \sum_{j=1}^{N} s(j\alpha) \Big| \leq c \log^2 N \, .$$

4) Use the Erdős–Turán inequality to prove (see Definition 7.1)

$$D_N \leq cN^{1/3} \left(\sum_{h=1}^{+\infty} \left| \frac{1}{h} \sum_{j=1}^{N} e^{2\pi i h \omega(j)} \right|^2 \right)^{1/3}.$$

5) Let p be an odd prime. Use the results in Chapter 4 to prove that

$$D_p \leq c\, p^{1/2} \log p.$$

8

Integer points and Poisson summation formula

In Chapter 2 we proved the following results (Theorems 2.19 and 2.14) for the arithmetic means of $r(n)$ and $d(n)$:

$$\frac{1}{R} \sum_{n \leq R} r(n) = \pi + O\left(R^{-1/2}\right) ,$$

$$\frac{1}{R} \sum_{n \leq R} d(n) = \log R + (2\gamma - 1) + O\left(R^{-1/2}\right) ,$$

as $R \to +\infty$, where γ is the Euler–Mascheroni constant. In the last chapters we shall use different techinques to improve these estimates and to study several related problems. We first need to introduce the Fourier integrals.

8.1 Fourier integrals

For every $f \in L^1(\mathbb{R}^d)$ we define its *Fourier transform*

$$\widehat{f}(\xi) := \int_{\mathbb{R}^d} f(t) e^{-2\pi i \xi \cdot t} \, dt ,$$

where $\xi \cdot t = \xi_1 t_1 + \xi_2 t_2 + \ldots + \xi_d t_d$ is the inner product in $\mathbb{R}^d$.

Proposition 8.1 *Let $f \in L^1(\mathbb{R}^d)$. Then the function $\widehat{f}(\xi)$ is uniformly continuous on $\mathbb{R}^d$.*

Proof Let $R > 0$ satisfy $\int_{|t|>R} |f(t)| \, dt \leq \varepsilon/4$. By (6.19) we have, for η small enough,

$$\left|\widehat{f}(\xi + \eta) - \widehat{f}(\xi)\right| = \left|\int_{\mathbb{R}^d} f(t) e^{-2\pi i(\xi+\eta)\cdot t} \, dt - \int_{\mathbb{R}^d} f(t) e^{-2\pi i \xi \cdot t} \, dt\right|$$

$$\leq \int_{\mathbb{R}^d} |f(t)| \left|e^{-2\pi i \eta \cdot t} - 1\right| \, dt = \int_{|t| \leq R} |f(t)| \left|e^{-2\pi i \eta \cdot t} - 1\right| \, dt + 2 \int_{|t|>R} |f(t)| \, dt$$

150

$$\leq 2\pi |\eta| \int_{|t| \leq R} |t| \, |f(t)| \, dt + \varepsilon/2 \leq \varepsilon .$$

$\square$

Theorem 8.2 *The Fourier transform and the differentiation are related as follows.*

(i) *Let $t = (t_1, \ldots, t_d) \in \mathbb{R}^d$ and $\xi = (\xi_1, \ldots, \xi_d) \in \mathbb{R}^d$. Assume that for a given k we have $\frac{\partial f}{\partial t_k} \in L^1(\mathbb{R}^d)$. Then*

$$\left(\frac{\partial f}{\partial t_k} \right)^{\wedge} (\xi) = 2\pi i \xi_k \, \widehat{f}(\xi) .$$

(ii) *Let $f \in L^1(\mathbb{R}^d)$ and assume that for a given k we have $t_k f(t) \in L^1(\mathbb{R}^d)$. Then $\widehat{f}(\xi)$ can be differentiated with respect to ξ_k and*

$$\frac{\partial \widehat{f}}{\partial \xi_k}(\xi) = (-2\pi i t_k \, f(t))^{\wedge}(\xi) .$$

Proof For the proof of (i) we refer to Lemma 7.13. In order to prove (ii), let $h = (0, \ldots, 0, h_k, 0, \ldots, 0)$. We apply the dominated convergence theorem to obtain

$$\frac{\widehat{f}(\xi + h) - \widehat{f}(\xi)}{h_k} = \frac{1}{h_k} \left(\int_{\mathbb{R}^d} f(t) \, e^{-2\pi i t \cdot (\xi + h)} \, dt - \int_{\mathbb{R}^d} f(t) \, e^{-2\pi i t \cdot \xi} \, dt \right)$$

$$= \frac{1}{h_k} \int_{\mathbb{R}^d} f(t) \, e^{-2\pi i t \cdot \xi} \left(e^{-2\pi i t \cdot h} - 1 \right) dt$$

$$= \left(\frac{e^{-2\pi i t_k h_k} - 1}{h_k} f(t) \right)^{\wedge} (\xi) \to (-2\pi i t_k f(t))^{\wedge}(\xi)$$

as $h_k \to 0$.

$\square$

Remark 8.3 Let S be the set of functions $\varphi \in C^\infty(\mathbb{R}^d)$ such that for every polynomial P on $\mathbb{R}^d$ and every choice of non-negative integers $\alpha_1, \ldots, \alpha_d$ we have

$$\sup_{t \in \mathbb{R}^d} \left| P(t) \left(\frac{\partial}{\partial t_1} \right)^{\alpha_1} \cdots \left(\frac{\partial}{\partial t_d} \right)^{\alpha_d} \varphi(t) \right| < \infty .$$

The previous result shows that if $\varphi \in S$, then $\widehat{\varphi} \in S$. We observe that S is dense in $L^p(\mathbb{R}^d)$ for every $1 \leq p < +\infty$.

The *convolution* is defined as on $\mathbb{T}^d$:

$$(f * g)(t) := \int_{\mathbb{R}^d} f(t - s)g(s) \, ds .$$

The proof of the following result is close to the corresponding argument for the periodic case, see Proposition 6.10 and Theorem 6.9.

Theorem 8.4   *We have*

(i) *If $f, g \in L^1(\mathbb{R}^d)$, then*

$$(f * g)^\wedge (\xi) = \widehat{f}(\xi)\, \widehat{g}(\xi) \, .$$

(ii) *If $f \in L^1(\mathbb{R}^d)$ and $g \in L^p(\mathbb{R}^d)$, $p \geq 1$, then*

$$\|f * g\|_{L^p(\mathbb{R}^d)} \leq \|f\|_{L^1(\mathbb{R}^d)}\, \|g\|_{L^p(\mathbb{R}^d)} \, .$$

(iii) *Let f be a measurable and non-negative function on $\mathbb{R}^d \times \mathbb{R}^d$, and let $1 \leq p < \infty$. Then we have the Minkowski integral inequality*

$$\left\{ \int_{\mathbb{R}^d} \left(\int_{\mathbb{R}^d} f(t, y)\, dy \right)^p dt \right\}^{1/p} \leq \int_{\mathbb{R}^d} \left\{ \int_{\mathbb{R}^d} f^p(t, y)\, dt \right\}^{1/p} dy \, .$$

Lemma 8.5   *Let $\phi \in L^1(\mathbb{R}^d)$ satisfy $\int_{\mathbb{R}^d} \phi = 1$. For every $\varepsilon > 0$ let*

$$\phi_\varepsilon(t) := \varepsilon^{-d} \phi(\varepsilon^{-1} t) \, . \tag{8.1}$$

Then, as $\varepsilon \to 0$,

(i) *if $1 \leq p < +\infty$ and $f \in L^p(\mathbb{R}^d)$, then $f * \phi_\varepsilon \to f$ in the L^p norm;*
(ii) *if f is bounded and continuous on $\mathbb{R}^d$, then $f * \phi_\varepsilon \to f$ uniformly.*

Proof   We prove only (i). Since $\int_{\mathbb{R}^d} \phi_\varepsilon = 1$ for every $\varepsilon > 0$, Minkowski's integral inequality yields

$$\left\{ \int_{\mathbb{R}^d} |(f * \phi_\varepsilon)(t) - f(t)|^p \, dt \right\}^{1/p}$$

$$= \left\{ \int_{\mathbb{R}^d} \left| \int_{\mathbb{R}^d} [f(t - s) - f(t)]\, \phi_\varepsilon(s)\, ds \right|^p dt \right\}^{1/p}$$

$$= \left\{ \int_{\mathbb{R}^d} \left| \varepsilon^{-d} \int_{\mathbb{R}^d} [f(t - s) - f(t)]\, \phi(\varepsilon^{-1} s)\, ds \right|^p dt \right\}^{1/p}$$

$$= \left\{ \int_{\mathbb{R}^d} \left| \int_{\mathbb{R}^d} [f(t - \varepsilon u) - f(t)]\, \phi(u)\, du \right|^p dt \right\}^{1/p}$$

$$\leq \int_{\mathbb{R}^d} \left\{ \int_{\mathbb{R}^d} |[f(t - \varepsilon u) - f(t)]\, \phi(u)|^p \, dt \right\}^{1/p} du$$

$$= \int_{\mathbb{R}^d} |\phi(u)| \left\{ \int_{\mathbb{R}^d} |f(t - \varepsilon u) - f(t)|^p \, dt \right\}^{1/p} du \, .$$

Then, by the dominated convergence theorem it is enough to prove that

$$\int_{\mathbb{R}^d} |f(t - v) - f(t)|^p \, dt \to 0 \tag{8.2}$$

as $v \to 0$. Indeed, (8.2) is true if f is continuous and has compact support. The general case follows by approximation. □

Lemma 8.6 *Let* $f(t) = e^{-\pi a|t|^2}$, *with* $t \in \mathbb{R}^d$, $a > 0$. *Then*

$$\widehat{f}(\xi) = a^{-d/2} e^{-\pi|\xi|^2/a} .$$

Proof We prove the 1-dimensional case first. Let $f(x) = e^{-\pi a x^2}$. By differentiation under the integral sign and integration by parts we obtain

$$\frac{d}{du}\widehat{f}(u) = (-2\pi i x f(x))^{\wedge}(u) = \int_{\mathbb{R}} (-2\pi i x) e^{-\pi a x^2} e^{-2\pi i u x} \, dx$$

$$= \frac{i}{a} \int_{\mathbb{R}} (-2\pi a x) e^{-\pi a x^2} e^{-2\pi i u x} \, dx = -\frac{i}{a} \int_{\mathbb{R}} e^{-\pi a x^2} e^{-2\pi i u x} (-2\pi i u) \, dx$$

$$= \frac{-2\pi u}{a} \widehat{f}(u) .$$

The differential equation

$$\frac{d}{du}\widehat{f}(u) = \frac{-2\pi u}{a} \widehat{f}(u)$$

has solution $\widehat{f}(u) = K e^{-\pi u^2/a}$, where, by (6.25),

$$K = \widehat{f}(0) = \int_{-\infty}^{\infty} e^{-\pi a x^2} \, dx = a^{-1/2} .$$

Now the proof of the d-dimensional case is simple:

$$\widehat{f}(\xi) = \int_{\mathbb{R}^d} e^{-\pi a|t|^2} e^{-2\pi i t \cdot \xi} \, dt$$

$$= \int_{\mathbb{R}} e^{-\pi a t_1^2} e^{-2\pi i t_1 \xi_1} \, dt_1 \int_{\mathbb{R}} e^{-\pi a t_2^2} e^{-2\pi i t_2 \xi_2} \, dt_2 \cdots \int_{\mathbb{R}} e^{-\pi a t_d^2} e^{-2\pi i t_d \xi_d} \, dt_d$$

$$= a^{-1/2} e^{-\pi \xi_1^2/a} \cdot a^{-1/2} e^{-\pi \xi_2^2/a} \cdots a^{-1/2} e^{-\pi \xi_d^2/a} = a^{-d/2} e^{-\pi|\xi|^2/a} .$$

□

We can now prove the *Fourier inversion formula*.

Theorem 8.7 (Fourier inversion formula) *Let* $f \in L^1(\mathbb{R}^d)$ *satisfy* $\widehat{f} \in L^1(\mathbb{R}^d)$. *Then* f *is (almost everywhere equal to) a continuous function and it is the inverse Fourier transform of its Fourier transform. That is*

$$f(t) = \int_{\mathbb{R}^d} \widehat{f}(\xi) e^{2\pi i \xi \cdot t} \, d\xi . \tag{8.3}$$

Proof For $t \in \mathbb{R}^d$ and $\varepsilon > 0$ let

$$\phi(\xi) = e^{2\pi i \xi \cdot t - \pi \varepsilon^2|\xi|^2} .$$

We apply the previous lemma, following the notation in (8.1),

$$\widehat{\phi}(y) = \int_{\mathbb{R}^d} e^{2\pi i \xi \cdot t - \pi \varepsilon^2 |\xi|^2} e^{-2\pi i \xi \cdot y}\, d\xi = \int_{\mathbb{R}^d} e^{-\pi \varepsilon^2 |\xi|^2} e^{-2\pi i \xi \cdot (y-t)}\, d\xi$$

$$= \varepsilon^{-d} e^{-\pi |y-t|^2/\varepsilon^2} = h_\varepsilon(y-t),$$

where $h(y) = e^{-\pi |y|^2}$. Since $\int_{\mathbb{R}^d} h = 1$, Lemma 8.5 implies, as $\varepsilon \to 0$,

$$\int_{\mathbb{R}^d} e^{-\pi \varepsilon^2 |\xi|^2} e^{2\pi i \xi \cdot t} \widehat{f}(\xi)\, d\xi = \int_{\mathbb{R}^d} \phi \widehat{f} = \int_{\mathbb{R}^d} \widehat{\phi} f = \int_{\mathbb{R}^d} f(y) h_\varepsilon(y-t)\, dt$$

$$= (f * h_\varepsilon)(t) \to f(t)$$

in $L^1\left(\mathbb{R}^d\right)$. Then there exists a sequence $\varepsilon_j \to 0$ such that, as $j \to +\infty$,

$$\int_{\mathbb{R}^d} e^{-\pi \varepsilon_j^2 |\xi|^2} e^{2\pi i \xi \cdot t} \widehat{f}(\xi) d\xi \longrightarrow f(t) \qquad \text{a.e.}$$

Since $\widehat{f} \in L^1\left(\mathbb{R}^d\right)$, the dominated convergence theorem implies

$$\int_{\mathbb{R}^d} e^{-\pi \varepsilon^2 |\xi|^2} e^{2\pi i \xi \cdot t} \widehat{f}(\xi)\, d\xi \longrightarrow \int_{\mathbb{R}^d} e^{2\pi i \xi \cdot t} \widehat{f}(\xi)\, d\xi$$

as $\varepsilon \to 0$. Then, up to a set of measure zero,

$$f(t) = \int_{\mathbb{R}^d} e^{2\pi i \xi \cdot t} \widehat{f}(\xi)\, d\xi.$$

The function $f(t)$ is continuous because it is the Fourier transform of an integrable function. □

Corollary 8.8   Let $f \in X := \left\{ f \in L^1(\mathbb{R}^d) : \widehat{f} \in L^1(\mathbb{R}^d) \right\}$. *Then f and $\widehat{f}$ belong to $L^2\left(\mathbb{R}^d\right)$ and*

$$\|f\|_{L^2(\mathbb{R}^d)} = \left\| \widehat{f} \right\|_{L^2(\mathbb{R}^d)}.$$

Proof   By the inversion formula (Theorem 8.7) we have $f \in L^\infty(\mathbb{R}^d)$. So

$$\int_{\mathbb{R}^d} |f|^2 \leq \|f\|_{L^\infty(\mathbb{R}^d)} \int_{\mathbb{R}^d} |f|.$$

Hence $f \in L^2\left(\mathbb{R}^d\right)$. Given $f, g \in X$, let $h = \overline{\overline{g}}$. Then Theorem 8.7 implies

$$\widehat{h}(y) = \int_{\mathbb{R}^d} \overline{\widehat{g}(\xi)} e^{-2\pi i y \cdot \xi}\, d\xi = \overline{\int_{\mathbb{R}^d} \widehat{g}(\xi) e^{2\pi i y \cdot \xi}\, d\xi} = \overline{g(y)},$$

and therefore

$$\int_{\mathbb{R}^d} f\, \overline{g} = \int_{\mathbb{R}^d} f\, \widehat{h} = \int_{\mathbb{R}^d} \widehat{f}\, h = \int_{\mathbb{R}^d} \widehat{f}\, \overline{\widehat{g}}.$$

If we let $g = f$ we obtain $\|f\|_{L^2(\mathbb{R}^d)} = \left\|\widehat{f}\right\|_{L^2(\mathbb{R}^d)}$ for every function in X. □

The density of S allows us to extend the previous result to every function $f \in L^1(\mathbb{R}^d) \cap L^2(\mathbb{R}^d)$. The following general result was proved by Plancherel in 1910 [141]. See, for example, [82] for a proof.

Theorem 8.9 (Plancherel) *If $f \in L^1(\mathbb{R}^d) \cap L^2(\mathbb{R}^d)$, then $\widehat{f} \in L^2(\mathbb{R}^d)$ and*

$$\|f\|_{L^2(\mathbb{R}^d)} = \left\|\widehat{f}\right\|_{L^2(\mathbb{R}^d)} . \tag{8.4}$$

Moreover, the map $\mathcal{F} : f \to \widehat{f}$ extends uniquely to a unitary isomorphism on $L^2\left(\mathbb{R}^d\right)$.

Now we start studying the behaviour of $\widehat{f}(\xi)$ for large $|\xi|$. Of course we have the easy bound

$$\left|\widehat{f}(\xi)\right| \leq \|f\|_{L^1(\mathbb{R}^d)} . \tag{8.5}$$

The main result is the Riemann–Lebesgue lemma.

Lemma 8.10 (Riemann–Lebesgue) *For every $f \in L^1(\mathbb{R}^d)$ we have $\widehat{f}(\xi) \to 0$ as $|\xi| \to \infty$.*

Proof A separation of variables and (7.7) imply the result for the characteristic function of a d-dimensional interval. Then the result is true for every finite linear combination of characteristic functions of d-dimensional intervals. Hence it is true for a dense subspace of $L^1(\mathbb{R}^d)$. Then for every $f \in L^1(\mathbb{R}^d)$ there exists a sequence $f_n \in L^1(\mathbb{R}^d)$ such that $\widehat{f_n}(\xi) \to 0$ as $|\xi| \to +\infty$ and $\|f - f_n\|_{L^1(\mathbb{R}^d)} \to 0$. Then

$$\sup_\xi \left|\widehat{f}(\xi) - \widehat{f_n}(\xi)\right| \leq \|f - f_n\|_{L^1(\mathbb{R}^d)} \longrightarrow 0$$

as $n \to \infty$. Hence $\widehat{f}(\xi) \to 0$. □

For certain families of functions we can say more on the decay of the Fourier transform. We need a lemma on convex functions.

Lemma 8.11 *Let f be a convex function on an open interval (a, b). Then f' exists on (a, b) except at most in a countable set. Moreover, f' is increasing.*

Proof The function $s \longmapsto \frac{f(x+s)-f(x)}{s}$ increases with s. Then the function $x \longmapsto f'_+(x)$ is increasing on (a, b). Indeed, if $x < y$,

$$f'_+(x) := \lim_{s \to 0^+} \frac{f(x+s) - f(x)}{s} \leq \frac{f(y) - f(x)}{y - x}$$

$$\leq \lim_{s \to 0^-} \frac{f(y+s) - f(y)}{s} := f'_-(y) \leq \lim_{s \to 0^+} \frac{f(y+s) - f(y)}{s} := f'_+(y) \, .$$

Then $f'_+(x)$ has at most a countable number of discontinuities. Since the continuity of f'_+ at x implies the continuity of f' at x, we end the proof. $\square$

The following two results are due to Podkorytov [142]. See also [35, 38].

Lemma 8.12 (Podkorytov) *Assume that $f : \mathbb{R} \to [0, +\infty)$ is supported and concave on the interval $[-1, 1]$. Then, for $|\xi| \geq 1$, we have*

$$\left| \widehat{f}(\xi) \right| = \left| \int_{-1}^{1} f(x) \, e^{-2\pi i \xi x} \, dx \right| \leq \frac{1}{|\xi|} \left(f\left(1 - \frac{1}{2|\xi|}\right) + f\left(-1 + \frac{1}{2|\xi|}\right) \right) . \quad (8.6)$$

Proof We may assume that $\xi \geq 1$. By Lemma 8.11 we can integrate by parts and obtain

$$\left| \widehat{f}(\xi) \right| \leq \frac{1}{2\pi \xi} f(1^-) + \frac{1}{2\pi \xi} f(-1^+) + \frac{1}{2\pi \xi} \left| \int_{-1}^{1} f'(x) \, e^{-2\pi i \xi x} \, dx \right| . \quad (8.7)$$

Assume that $f(\alpha) \geq f(x)$ for every $x \in [-1, 1]$. Then f increases in $[-1, \alpha]$ and decreases in $[\alpha, 1]$. We can assume that $0 \leq \alpha \leq 1$. Then $f(-1^+) \leq f(-1 + 1/(2\xi))$. In order to estimate $f(1^-)$ we observe that when $\alpha \leq 1 - 1/(2\xi)$ we have $f(1^-) \leq f(1 - 1/(2\xi))$. Since f is concave in the interval $[-1, 1]$, if $\alpha > 1 - 1/(2\xi)$ we have

$$f(1^-) \leq f(\alpha) \leq 2f(0) \leq 2f(1 - 1/(2\xi)) \, .$$

In order to estimate the integral in (8.7) we change variables,

$$I := \int_{-1}^{1} f'(x) \, e^{-2\pi i \xi x} \, dx = - \int_{-1 + \frac{1}{2\xi}}^{1 + \frac{1}{2\xi}} f'\left(x - \frac{1}{2\xi}\right) e^{-2\pi i \xi x} \, dx \, .$$

Hence

$$2I = \int_{-1}^{1} f'(x) \, e^{-2\pi i \xi x} \, dx - \int_{-1 + \frac{1}{2\xi}}^{1 + \frac{1}{2\xi}} f'\left(x - \frac{1}{2\xi}\right) e^{-2\pi i \xi x} \, dx$$

$$= \int_{-1}^{-1 + \frac{1}{2\xi}} f'(x) \, e^{-2\pi i \xi x} \, dx + \int_{-1 + \frac{1}{2\xi}}^{1} \left(f'(x) - f'\left(x - \frac{1}{2\xi}\right) \right) e^{-2\pi i \xi x} \, dx$$

$$+ \int_{1}^{1 + \frac{1}{2\xi}} f'\left(x - \frac{1}{2\xi}\right) e^{-2\pi i \xi x} \, dx$$

$$:= I_1 + I_2 + I_3 \, .$$

Since $0 \leq \alpha \leq 1$ we have

$$|I_1| \leq \int_{-1}^{-1 + \frac{1}{2\xi}} f'(x) \, dx = f\left(-1 + \frac{1}{2\xi}\right) - f(-1^+) \leq f\left(-1 + \frac{1}{2\xi}\right) .$$

The estimate for I_3 is similar when $\alpha \le 1 - 1/(2\xi)$. If $\alpha > 1 - 1/(2\xi)$, then

$$|I_3| \le \int_1^{\alpha+\frac{1}{2\xi}} f'\left(x - \frac{1}{2\xi}\right) dx - \int_{\alpha+\frac{1}{2\xi}}^{1+\frac{1}{2\xi}} f'\left(x - \frac{1}{2\xi}\right) dx$$

$$= 2f(\alpha) - f\left(1 - \frac{1}{2\xi}\right) - f(1^-) \le 2f(\alpha) \le 4f(0) \le 4f\left(1 - \frac{1}{2\xi}\right).$$

As for I_2, the monotonicity of f' implies

$$|I_2| \le \int_{-1+\frac{1}{2\xi}}^1 \left[f'\left(x - \frac{1}{2\xi}\right) - f'(x) \right] dx$$

$$= f\left(1 - \frac{1}{2\xi}\right) - f(-1^+) - f(1^-) + f\left(-1 + \frac{1}{2\xi}\right)$$

$$\le f\left(1 - \frac{1}{2\xi}\right) + f\left(-1 + \frac{1}{2\xi}\right).$$

$\square$

The previous lemma has the following geometric meaning. Assume for simplicity that $f(x)$ is even. Then

$$\int_{-1}^1 f'(x) e^{-2\pi i \xi x} dx = -i \int_{-1}^1 f'(x) \sin(2\pi\xi x) \, dx.$$

Let $\xi \ge 1$. In the figure we overlap the graphs of $f'(x)$ and $f'(x) \sin(2\pi\xi x)$ on $[0, 1]$:

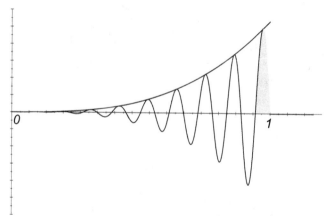

Then we see that $\int_{-1}^1 f'(x) \sin(2\pi\xi x) \, dx$ behaves like a Leibniz sum. Indeed, it is (from right to left) a sum of terms having alternating signs and decreasing absolute value. Then it is smaller than the first term, that is, the area of the shaded part, which is essentially contained in a rectangle with sides $1/\xi$ and

$f'(s - 1/(2\xi))$. We acknowledge that the above figure represents a suitable choice of ξ.

We deduce a useful geometric estimate for the Fourier transform of the characteristic function of a planar convex body.

Theorem 8.13 (Podkorytov) *Let $C \subset \mathbb{R}^2$ be a convex body. We write $\Theta := (\cos\theta, \sin\theta)$ and, for $0 \le \theta < \pi$ and small $\delta > 0$, let*

$$\lambda(\delta, \theta) = \lambda_C(\delta, \theta) := \left\{ t \in C : \delta + t \cdot \Theta = \sup_{y \in C} (y \cdot \Theta) \right\} \tag{8.8}$$

be the chord perpendicular to Θ and 'at distance δ from the boundary' ∂C of C (see the following figure). Then, for large $\rho > 0$ we have

$$\left| \widehat{\chi_C}(\rho\Theta) \right| \le c\rho^{-1} \left(\left| \lambda_C(\rho^{-1}, \theta) \right| + \left| \lambda_C(\rho^{-1}, \theta + \pi) \right| \right) , \tag{8.9}$$

where $|\lambda_C|$ is the length of the segment λ_C.

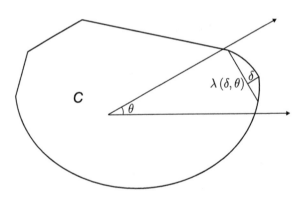

Proof   We may assume that $\Theta = (1, 0)$. Then

$$\widehat{\chi_C}(\xi_1, 0) = \int_{-\infty}^{+\infty} \left(\int_{-\infty}^{+\infty} \chi_C(t_1, t_2) \, dt_2 \right) e^{-2\pi i \xi_1 t_1} \, dt_1 = \widehat{h}(\xi_1) ,$$

where $h(s)$ is the length of the intersection of C with the line $t_1 = s$. Let $[a, b]$ be the support of $h(s)$. Observe that $h(s)$ can be seen as the difference between a concave function and a convex function on $[a, b]$. Hence $h(s)$ is concave on [a, b] and the previous lemma[1] implies

$$\left|\hat{h}(\xi_1)\right| \le \frac{c}{|\xi_1|} \left[h\left(b - \frac{1}{2|\xi_1|}\right) + h\left(a + \frac{1}{2|\xi_1|}\right) \right]$$

$$\le c\rho^{-1} \left(\left|\lambda_C(\rho^{-1}, 0)\right| + \left|\lambda_C(\rho^{-1}, \pi)\right| \right) .$$

□

Let C be any convex planar body. The previous lemma yields, for every $\xi \in \mathbb{R}^2$,

$$\left|\hat{\chi}_C(\xi)\right| \le \frac{c}{1 + |\xi|} . \tag{8.10}$$

Observe that (8.10) cannot be improved. Indeed, for a square $Q = [-1/2, 1/2]^2$ we have

$$\hat{\chi}_Q(\xi_1, \xi_2) = \int_{-1/2}^{1/2} \int_{-1/2}^{1/2} e^{-2\pi i(t_1\xi_1 + t_2\xi_2)} \, dt_1 dt_2 = \frac{\sin(\pi\xi_1)}{\pi\xi_1} \frac{\sin(\pi\xi_2)}{\pi\xi_2} \tag{8.11}$$

(with obvious modifications when ξ_1 or ξ_2 vanish) and therefore, for every integer n,

$$\hat{\chi}_Q\left(2n + \frac{1}{2}, 0\right) = \frac{1}{\pi(2n + 1/2)} .$$

The next lemma shows that the case of a disc is different.

Lemma 8.14 Let $\chi_1(t) := \chi_{B(0,1)}(t)$ *be the characteristic function of the disc* $B(0, 1) \subset \mathbb{R}^2$ *having centre 0 and radius 1. Then there exists a positive constant* c *such that, for every* $\xi \in \mathbb{R}^2$,

$$\left|\hat{\chi}_1(\xi)\right| \le \frac{c}{1 + |\xi|^{3/2}} .$$

Proof We observe that the length of the chords at distance $|\xi|^{-1}$ from the boundary is $\approx |\xi|^{-1/2}$. Then we apply (8.5) and Lemma 8.12. □

In the last chapter we shall prove that, as $|\xi| \to \infty$,

$$\hat{\chi}_1(\xi) = \frac{1}{\pi} |\xi|^{-3/2} \cos\left(2\pi|\xi| - \frac{3\pi}{4}\right) + O(|\xi|^{-5/2}) .$$

---

[1] We first pass from $[a, b]$ to the symmetric support $[-\alpha, \alpha]$. Then we define $k(x) = h(\alpha x)$, so that

$$\left|\hat{h}(\xi)\right| = \left| \int_{-\alpha}^{\alpha} h(s)e^{-2\pi i s\xi} ds \right| = \alpha \left| \int_{-1}^{1} h(\alpha t)e^{-2\pi i\xi\alpha t} dt \right| = \alpha \left|\hat{k}(\alpha\xi)\right|$$

$$\le \alpha \left| k\left(1 - \frac{1}{2|\alpha\xi|}\right) + k\left(-1 + \frac{1}{2|\alpha\xi|}\right) \right| = \alpha \left| h\left(\alpha - \frac{1}{2|\xi|}\right) + h\left(-\alpha + \frac{1}{2|\xi|}\right) \right| .$$

8.2 The Poisson summation formula

The Poisson summation formula is a bridge between Fourier integrals and Fourier series. The basic idea is to consider a function $f \in L^1\left(\mathbb{R}^d\right)$ and periodize it.

Theorem 8.15 *Let $f \in L^1(\mathbb{R}^d)$. Then there exists $g \in L^1(\mathbb{T}^d)$ such that*

$$\|g\|_{L^1(\mathbb{T}^d)} \le \|f\|_{L^1(\mathbb{R}^d)} , \qquad \widehat{g}(m) = \widehat{f}(m) \qquad \text{for every } m \in \mathbb{Z}^d .$$

Observe that $\widehat{f}(m)$ is a Fourier transform of the function $f \in L^1(\mathbb{R}^d)$, while $\widehat{g}(m)$ is a Fourier coefficient of the periodic function $g \in L^1(\mathbb{T}^d)$.

Proof The function

$$g(t) = \sum_{k \in \mathbb{Z}^d} f(t + k)$$

is periodic (that is, $g(t + k) = g(t)$ for every $t \in \mathbb{R}^d$ and $k \in \mathbb{Z}^d$). We have

$$\|g\|_{L^1(\mathbb{T}^d)} = \int_{[-\frac{1}{2},\frac{1}{2})^d} |g(t)| \, dt \le \sum_{k \in \mathbb{Z}^d} \int_{[-\frac{1}{2},\frac{1}{2})^d} |f(t + k)| \, dt$$

$$= \sum_{k \in \mathbb{Z}^d} \int_{[-\frac{1}{2},\frac{1}{2})^d - k} |f(t)| \, dt = \int_{\mathbb{R}^d} |f(t)| \, dt .$$

By the dominated convergence theorem we have, for every $m \in \mathbb{Z}^d$,

$$\widehat{g}(m) = \int_{[-\frac{1}{2},\frac{1}{2})^d} \sum_{k \in \mathbb{Z}^d} f(t + k) \, e^{-2\pi i m \cdot t} \, dt = \sum_{k \in \mathbb{Z}^d} \int_{[-\frac{1}{2},\frac{1}{2})^d} f(t + k) \, e^{-2\pi i m \cdot t} \, dt$$

$$= \sum_{k \in \mathbb{Z}^d} \int_{[-\frac{1}{2},\frac{1}{2})^d - k} f(t) \, e^{-2\pi i m \cdot t} \, dt = \int_{\mathbb{R}^d} f(t) \, e^{-2\pi i m \cdot t} \, dt = \widehat{f}(m) .$$

$\square$

The following result was proved by Poisson between 1823 and 1827.

Theorem 8.16 (Poisson summation formula) *Let $f \in L^1(\mathbb{R}^d)$ and let $a, c > 0$ satisfy*

$$|f(t)| \le \frac{c}{(1 + |t|)^{d+a}} , \qquad \left|\widehat{f}(\xi)\right| \le \frac{c}{(1 + |\xi|)^{d+a}} \tag{8.12}$$

for every $t, \xi \in \mathbb{R}^d$. Then f is (a.e. equal to) a continuous function and for every $t \in \mathbb{T}^d$ we have

$$\sum_{m \in \mathbb{Z}^d} f(t + m) = \sum_{m \in \mathbb{Z}^d} \widehat{f}(m) \, e^{2\pi i m \cdot t} . \tag{8.13}$$

Letting t = 0 we obtain

$$\sum_{m \in \mathbb{Z}^d} f(m) = \sum_{m \in \mathbb{Z}^d} \widehat{f}(m) .$$

(8.14)

The particular case (8.14) is very elegant and is usually termed the Poisson summation formula.

Proof Let $g(t) = \sum_{m \in \mathbb{Z}^d} f(t + m)$. By (8.12) and Theorem 8.15 we have $g \in L^1(\mathbb{T}^d)$ and $\widehat{g}(m) = \widehat{f}(m)$ for every $m \in \mathbb{Z}^d$. The second inequality in (8.12) implies

$$\sum_{m \in \mathbb{Z}^d} |\widehat{g}(m)| \le c \int_0^{+\infty} \frac{\rho^{d-1}}{(1 + \rho)^{d+a}} \, d\rho < +\infty ,$$

so that the series $\sum_{m \in \mathbb{Z}^d} \widehat{g}(m) \, e^{2\pi i m \cdot t}$ converges absolutely and uniformly. Then

$$\sum_{m \in \mathbb{Z}^d} f(t + m) = \sum_{m \in \mathbb{Z}^d} \widehat{f}(m) \, e^{2\pi i m \cdot t}$$

for every $t \in \mathbb{T}^d$. □

As an exercise we deduce, for every $x \notin \mathbb{Z}$, the identity

$$\sum_{n=-\infty}^{+\infty} \frac{1}{(x + n)^2} = \frac{\pi^2}{\sin^2(\pi x)} .$$

(8.15)

Indeed, the function

$$A(y) := \begin{cases} 1 - |y| & \text{if } |y| \le 1, \\ 0 & \text{if } |y| > 1 \end{cases}$$

satisfies, when $\eta \ne 0$,

$$\widehat{A}(\eta) = \int_{-1}^{1} (1 - |y|) \, e^{-2\pi i \eta y} \, dy = 2 \int_0^1 (1 - y) \cos(2\pi \eta y) \, dy = \frac{\sin^2(\pi \eta)}{\pi^2 \eta^2} .$$

By the Fourier inversion formula (Theorem 8.7), the function

$$\phi(x) := \left(\frac{\sin(\pi x)}{\pi x} \right)^2$$

has Fourier transform

$$\widehat{\phi}(\xi) = \begin{cases} 1 - |\xi| & \text{if } |\xi| \le 1, \\ 0 & \text{if } |\xi| > 1. \end{cases}$$

(8.16)

We apply the Poisson summation formula (8.13) to $\phi(x)$. By (8.16) we obtain

$$\sum_{n \in \mathbb{Z}} \left(\frac{\sin(\pi(x + n))}{\pi(x + n)} \right)^2 = \sum_{n \in \mathbb{Z}} \widehat{\phi}(n) \, e^{2\pi i n x} = \widehat{\phi}(0) = 1 .$$

This implies (8.15).

We can use (8.15) to obtain another proof of (4.35). Indeed,

$$\sum_{n=0}^{+\infty} \frac{1}{(2n+1)^2} = \frac{1}{4} \sum_{n=0}^{+\infty} \frac{1}{(n+1/2)^2} = \frac{1}{8} \sum_{n=-\infty}^{+\infty} \frac{1}{(n+1/2)^2} \qquad (8.17)$$

$$= \frac{1}{8} \left(\frac{\pi}{\sin(\pi/2)} \right)^2 = \frac{\pi^2}{8} .$$

Let $X := \sum_{n=1}^{+\infty} n^{-2}$, then we obtain $X/4 := \sum_{n=1}^{+\infty} (2n)^{-2}$, so that, by (8.17), $X - X/4 = \pi^2/8$. Hence $X = \pi^2/6$.

The Poisson summation formula is related to the Euler–Maclaurin summation formula (Lemma 6.17). See [113, p. 22].

8.3 The Gauss circle problem

For every $R > 0$ let $\chi_R(t) := \chi_{B(0,R)}(t)$ be the characteristic function of the disc

$$B_R := B(0,R) = \left\{ t \in \mathbb{R}^2 : |t| \leq R \right\}$$

with centre 0 and radius R. Let R be large and let

$$N(R) := \text{card}\left(B_R \cap \mathbb{Z}^2 \right) = \sum_{m \in \mathbb{Z}^2} \chi_R(m)$$

be the number of integer points in B_R. Let

$$D_R := N(R) - \pi R^2 .$$

We know (see (2.15)) that $D_R = O(R)$. The following result was proved by Sierpinski in 1906 [155] (see also [113, 160]).

Theorem 8.17 (Sierpinski)   *There exists $c > 0$ such that $|D_R| \leq cR^{2/3}$.*

Proof   First we use a convolution to smooth the discontinuous function $\chi_R(t)$. Let $\varepsilon > 0$ be small (we shall choose it later on) and let

$$\varphi_\varepsilon(t) := \pi^{-1} \varepsilon^{-2} \chi_\varepsilon(t) . \qquad (8.18)$$

The support of φ_ε is contained in $B(0,\varepsilon)$. Moreover, $\int_{\mathbb{R}^2} \varphi_\varepsilon = 1$ and

$$\widehat{\varphi_\varepsilon}(\xi) = \int_{\mathbb{R}^2} \pi^{-1} \varepsilon^{-2} \chi_\varepsilon(t) \, e^{-2\pi i \xi \cdot t} \, dt = \pi^{-1} \int_{\mathbb{R}^2} \chi_\varepsilon(t\varepsilon) \, e^{-2\pi i \xi \cdot t\varepsilon} \, dt \qquad (8.19)$$

$$= \pi^{-1} \int_{\mathbb{R}^2} \chi_1(t) \, e^{-2\pi i \varepsilon \xi \cdot t} \, dt = \pi^{-1} \widehat{\chi_1}(\varepsilon \xi) .$$

Let

$$\widetilde{\chi}_R^{(\varepsilon)}(t) := (\varphi_\varepsilon * \chi_R)(t), \qquad \widetilde{N}^\varepsilon(R) := \sum_{m \in \mathbb{Z}^2} \widetilde{\chi}_R^{(\varepsilon)}(m).$$

We are going to compare $N(R)$ with $\widetilde{N}^\varepsilon(R)$. By Lemma 8.14 the function $\widetilde{\chi}_R^{(\varepsilon)}(t)$ has absolutely convergent Fourier series, hence it is continuous. Moreover, it coincides with $\chi_R(t)$ when t is away from the boundary of the disc. Namely,

$$\chi_R(t) = \widetilde{\chi}_R^{(\varepsilon)}(t) \qquad \text{if } t \notin B_{R+\varepsilon} \backslash B_{R-\varepsilon}.$$

Indeed, if $t \notin B_{R+\varepsilon}$, then $|t - y| > \varepsilon$ for every $y \in B_R$ and, therefore,

$$\widetilde{\chi}_R^{(\varepsilon)}(t) = \int_{B_R} \varphi_\varepsilon(t - y)\, dy = 0 = \chi_R(t).$$

If $t \in B_{R-\varepsilon}$ then $B(t, \varepsilon) \subseteq B_R$ and, therefore,

$$\widetilde{\chi}_R^{(\varepsilon)}(t) = \int_{B_R} \varphi_\varepsilon(t - y)\, dy = \int_{B_\varepsilon} \varphi_\varepsilon(u)\, du = 1 = \chi_R(t).$$

Since, for every t,

$$0 \le \widetilde{\chi}_R^{(\varepsilon)}(t) = \int_{\mathbb{R}^2} \varphi_\varepsilon(t - y)\chi_R(y)\, dy \le \int_{\mathbb{R}^2} \varphi_\varepsilon(t - y)\, dy = \int_{\mathbb{R}^2} \varphi_\varepsilon = 1,$$

we have

$$\widetilde{N}^\varepsilon(R - \varepsilon) = \sum_{m \in \mathbb{Z}^2} \widetilde{\chi}_{R-\varepsilon}^{(\varepsilon)}(m) \le \sum_{m \in \mathbb{Z}^2} \chi_R(m) = N(R) \qquad (8.20)$$

$$\le \sum_{m \in \mathbb{Z}^2} \widetilde{\chi}_{R+\varepsilon}^{(\varepsilon)}(m) = \widetilde{N}^\varepsilon(R + \varepsilon).$$

Since

$$\widehat{\chi}_R(\xi) = \int_{B_R} e^{-2\pi i \xi \cdot t}\, dt = R^2 \int_{B_1} e^{-2\pi i \xi \cdot Rs}\, ds = R^2\, \widehat{\chi}_1(R\xi),$$

Theorem 8.4 and (8.19) give

$$\left(\widetilde{\chi}_R^{(\varepsilon)}\right)^{\wedge}(\xi) = \widehat{\varphi}_\varepsilon(\xi)\widehat{\chi}_R(\xi) = \left(\pi^{-1}\,\widehat{\chi}_1(\varepsilon\xi)\right)\left(R^2\,\widehat{\chi}_1(R\xi)\right). \qquad (8.21)$$

By Lemma 8.14, (8.19) and (8.21), we can apply the Poisson summation formula to the function $\widetilde{\chi}_R^{(\varepsilon)}(t)$ and obtain

$$\widetilde{N}^\varepsilon(R) = \sum_{m \in \mathbb{Z}^2} \widetilde{\chi}_R^{(\varepsilon)}(m) = \sum_{m \in \mathbb{Z}^2} \left(\widetilde{\chi}_R^{(\varepsilon)}\right)^{\wedge}(m) = \sum_{m \in \mathbb{Z}^2} \widehat{\varphi}_\varepsilon(m)\widehat{\chi}_R(m)$$

$$= \pi R^2 + \pi^{-1} R^2 \sum_{m \in \mathbb{Z}^2,\ m \ne 0} \widehat{\chi}_1(\varepsilon m)\widehat{\chi}_1(Rm).$$

We apply Lemma 8.14 again, then we bound the series with an integral and use polar coordinates:

$$R^2 \left| \sum_{m \in \mathbb{Z}^2,\, m \neq 0} \widehat{\chi_1}(\varepsilon m)\, \widehat{\chi_1}(Rm) \right| \leq c_1 R^2 \sum_{m \in \mathbb{Z}^2,\, m \neq 0} (1 + |\varepsilon m|)^{-3/2}\, |Rm|^{-3/2}$$

$$\leq c_2 R^{1/2} \int_{|\xi| \geq 1} \frac{1}{(1 + \varepsilon |\xi|)^{3/2} |\xi|^{3/2}}\, d\xi = c_2 R^{1/2} \int_1^\infty r^{-1/2} \frac{1}{(1 + \varepsilon r)^{3/2}}\, dr$$

$$\leq c_2 R^{1/2} \varepsilon^{-1/2} \int_0^\infty s^{-1/2} \frac{1}{(1 + s)^{3/2}}\, ds = c_3 R^{1/2} \varepsilon^{-1/2} \;.$$

Hence

$$\widetilde{N}^\varepsilon(R) = \pi R^2 + O\left(R^{1/2} \varepsilon^{-1/2}\right) \;. \tag{8.22}$$

Now we replace R with $R \pm \varepsilon$ in (8.22), then (8.20) implies

$$N(R) \leq \widetilde{N}^\varepsilon(R + \varepsilon) = \pi (R + \varepsilon)^2 + O\left((R + \varepsilon)^{1/2} \varepsilon^{-1/2}\right)$$
$$= \pi R^2 + 2\pi R\varepsilon + O\left(R^{1/2} \varepsilon^{-1/2}\right) = \pi R^2 + O(R\varepsilon + R^{1/2} \varepsilon^{-1/2})$$

together with a similar estimate from below. Let $\varepsilon = R^{-1/3}$ (this choice makes the terms $R\varepsilon$ and $R^{1/2} \varepsilon^{-1/2}$ equal). Then

$$N(R) = \pi R^2 + O(R^{2/3}) \;.$$

$\square$

Sierpinski's result is one step in a sequence of estimates for the circle problem. Let

$$\theta := \inf \left\{ \alpha \in \mathbb{R} : N(R) - \pi R^2 = O(R^\alpha) \right\} \;.$$

The following results have been obtained so far:

$\theta \leq 1$ Gauss (1801)

$\theta \leq 2/3$ Sierpinski (1906)

$\theta \leq 2/3 - \varepsilon$ van der Corput (1923)

$\theta \leq 37/56 = 0.66071 \cdots$ Landau (1924), Littlewood and Walfisz (1924)

$\theta \leq 163/247 = 0.65992 \cdots$ Walfisz (1927)

$\theta \leq 27/41 = 0.65854 \cdots$ Nieland (1928)

$\theta \leq 15/23 = 0.65217 \cdots$ Titchmarsh (1934)

$\theta \leq 13/20 = 0.65$ Hua (1942)

$\theta \leq 24/37 = 0.64865 \cdots$ Chen (1963)

$\theta \leq 35/54 = 0.64815 \cdots$ Nowak (1984)

$\theta \leq 278/429 = 0.64802 \cdots$ Kolesnik (1985)

$\theta \leq 7/11 = 0.63636 \cdots$ Iwaniec and Mozzochi (1987)

$\theta \leq 46/73 = 0.63014\cdots$ Huxley (1993)

$\theta \leq 131/208 = 0.62981\cdots$ Huxley (2003).

In 1915 Hardy and Landau proved independently that $\theta \geq 1/2$ [88, 115]. In a subsequent paper [89], Hardy proved that, as $R \to \infty$,

$$\frac{N(R) - \pi R^2}{\sqrt{R \log R}} \nrightarrow 0 , \tag{8.23}$$

which shows that the bound $\left| N(R) - \pi R^2 \right| \leq c\, R^{1/2}$ is false.

We present a proof of the result of Hardy and Landau which follows a general argument due to Erdős and Fuchs [75, 130], and has been communicated to us by Podkorytov.

Theorem 8.18 (Hardy–Landau)   *Let* $N(R) := \operatorname{card}\{\mathbb{Z}^2 \cap B(0,R)\}$ *and let* $D(R) := N(R) - \pi R^2$. *Assume the existence of two constants* $c_0 > 0$ *and* $\alpha < 1$ *such that* $|D(R)| \leq c_0 R^\alpha$ *for every* $R \geq 1$. *Then* $\alpha \geq 1/2$.

Proof   The power series

$$f(z) = \sum_{h=-\infty}^{+\infty} z^{h^2}$$

is defined on the open unit disc $\{z \in \mathbb{C} : |z| < 1\}$. As in Chapter 2, let $r(m) = \operatorname{card}\{(a,b) \in \mathbb{Z}^2 : a^2 + b^2 = m\}$. Then $r(m) = N\left(\sqrt{m}\right) - N\left(\sqrt{m-1}\right)$ for every positive integer m. Therefore

$$f^2(z) = \sum_{h,k=-\infty}^{+\infty} z^{h^2+k^2} = \sum_{m=0}^{+\infty} r(m) z^m \tag{8.24}$$

$$= 1 + \sum_{m=1}^{+\infty} \left(N\left(\sqrt{m}\right) - N\left(\sqrt{m-1}\right) \right) z^m = (1-z) \sum_{m=0}^{+\infty} N\left(\sqrt{m}\right) z^m$$

$$= (1-z) \sum_{m=0}^{+\infty} \left(D\left(\sqrt{m}\right) + \pi m \right) z^m = \frac{\pi z}{1-z} + (1-z) \sum_{m=0}^{+\infty} D\left(\sqrt{m}\right) z^m .$$

Since $\log x \leq x - 1$ for every $x \in \mathbb{R}$, then, for $0 < r < 1$, (6.25) implies

$$f(r) > \sum_{h=0}^{+\infty} r^{h^2} \geq \int_0^{+\infty} r^{t^2} dt = \frac{1}{\sqrt{\log(1/r)}} \int_0^{+\infty} e^{-u^2} du \tag{8.25}$$

$$= \frac{\sqrt{\pi}}{2\sqrt{\log(1/r)}} \geq \frac{\sqrt{\pi r}}{2\sqrt{1-r}} .$$

For every positive integer K let

$$S_K(z) := 1 + z + \ldots + z^{K-1} , \qquad I_K(r) := \int_{-\pi}^{\pi} \left| f\left(re^{i\theta}\right) S_K\left(re^{i\theta}\right) \right|^2 d\theta .$$

Observe that fS_K is a power series with non-negative integral coefficients. Let us write

$$(fS_K)(z) = \sum_{m=0}^{+\infty} \alpha_m z^m .$$

By Parseval's theorem and (8.25) we obtain

$$I_K(r) = 2\pi \int_0^1 \left| \sum \alpha_m r^m e^{2\pi i m\theta} \right|^2 d\theta = 2\pi \sum_{m=0}^{+\infty} \alpha_m^2 r^{2m} \geq 2\pi \sum_{m=0}^{+\infty} \alpha_m r^{2m}$$

$$= 2\pi f\left(r^2\right) S_K\left(r^2\right) \geq \frac{\pi^{3/2} r}{\sqrt{1-r^2}} K r^{2(K-1)} \geq \frac{K r^{2K}}{\sqrt{1-r}} .$$

We now want to bound $I_K(r)$ from above. (8.24) implies

$$I_K(r) \leq \int_{-\pi}^{\pi} \left| 1 + re^{i\theta} + \ldots + r^{K-1} e^{i(K-1)\theta} \right|^2 \left| \frac{\pi r e^{i\theta}}{1 - re^{i\theta}} \right| d\theta$$

$$+ \int_{-\pi}^{\pi} \left| 1 + re^{i\theta} + \ldots + r^{K-1} e^{i(K-1)\theta} \right|^2 \left| \left(1 - re^{i\theta}\right) \sum_{m=0}^{+\infty} D\left(\sqrt{m}\right) r^m e^{im\theta} \right| d\theta$$

$$:= A + B .$$

We estimate A. For every $\theta \in [-\pi, \pi]$ we have

$$\left| 1 - re^{i\theta} \right|^2 = \left(1 - re^{i\theta}\right)\left(1 - re^{-i\theta}\right) = (1-r)^2 + 2r(1 - \cos\theta)$$

$$= (1-r)^2 + 4r \sin^2\left(\frac{\theta}{2}\right) \geq (1-r)^2 + \frac{4r}{\pi^2}\theta^2 \geq \frac{1}{2}\left(\max\left(1 - r, \frac{2r}{\pi}\theta\right)\right)^2$$

(since $\sin t \geq \frac{2}{\pi} t$ for every $0 \leq t \leq \pi/2$). Then

$$A \leq \pi K^2 \int_{-\pi}^{\pi} \frac{1}{\left|1 - re^{i\theta}\right|} d\theta \leq c_1 K^2 \int_0^{\pi} \frac{1}{\max\left((1-r), \theta\right)} d\theta$$

$$\leq c_1 K^2 \frac{1}{1-r} \int_0^{1-r} d\theta + c_1 K^2 \int_{1-r}^{\pi} \theta^{-1} d\theta \leq c_2 K^2 \log\left(\frac{1}{1-r}\right) .$$

We estimate B. By the Cauchy–Schwarz inequality, Parseval's identity and the assumption $|D(R)| \leq c_0 R^\alpha$ we have

$$B \leq 2 \int_{-\pi}^{\pi} \left| 1 + re^{i\theta} + \ldots + r^{K-1} e^{i(K-1)\theta} \right| \left| \sum_{m=0}^{+\infty} D\left(\sqrt{m}\right) r^m e^{im\theta} \right| d\theta$$

$$\leq 2 \left\{ \int_{-\pi}^{\pi} \left| 1 + re^{i\theta} + \ldots + r^{K-1} e^{i(K-1)\theta} \right|^2 d\theta \right\}^{1/2}$$

$$\times \left\{ \int_{-\pi}^{\pi} \left| \sum_{m=0}^{+\infty} D\left(\sqrt{m}\right) r^m e^{im\theta} \right|^2 d\theta \right\}^{1/2}$$

$$\leq 4\pi \left\{ \sum_{m=0}^{K-1} r^{2m} \right\}^{1/2} \left\{ \sum_{m=0}^{+\infty} D^2\left(\sqrt{m}\right) r^{2m} \right\}^{1/2}$$

$$\leq 4\pi c_0 K^{1/2} \left\{ \sum_{m=0}^{+\infty} m^\alpha r^{2m} \right\}^{1/2}$$

$$\leq 4\pi c_0 K^{1/2} \left\{ \sum_{m=0}^{+\infty} m^\alpha r^m \right\}^{1/2}$$

$$\leq 4\pi c_0 \frac{K^{1/2}}{(1-r)^{(\alpha+1)/2}} \,,$$

where the last step is a consequence of Hölder's inequality. Indeed,[2] for $0 < a < 1$ and $0 < r < 1$ we have

$$\sum_{m=0}^{\infty} m^a r^m = \sum_{m=0}^{\infty} (m^a r^{am})(r^{m-am}) \leq \left\{ \sum_{m=0}^{\infty} mr^m \right\}^a \left\{ \sum_{m=0}^{\infty} r^m \right\}^{1-a}$$

$$\leq \left\{ \frac{1}{(1-r)^2} \right\}^a \left\{ \frac{1}{1-r} \right\}^{1-a} = \frac{1}{(1-r)^{a+1}} \,.$$

Hence

$$\frac{K r^{2K}}{(1-r)^{1/2}} \leq I_K(r) \leq c_2 K^2 \log\left(\frac{1}{1-r}\right) + 4\pi c_0 \frac{K^{1/2}}{(1-r)^{(\alpha+1)/2}} \,. \tag{8.26}$$

We now choose $r = r_K := 1 - K^{-b}$, with $b > 2$. Then (8.26) implies

$$c_3 K^{1+b/2} \left(1 - \frac{1}{K^b}\right)^{2K} \leq bK^2 \log K + K^{1/2} K^{b(\alpha+1)/2} \,.$$

Observe that $\left(1 - K^{-b}\right)^{2K} \to 1$ as $K \to +\infty$. Then for large K we have $\left(1 - K^{-b}\right)^{2K} \geq 1/2$. Hence there exists $c_4 > 0$ such that

$$c_4 K^{1+b/2} \leq bK^2 \log K + K^{1/2} K^{b(\alpha+1)/2} \,. \tag{8.27}$$

Since $b > 2$, then (8.27) implies

$$1 + \frac{b}{2} \leq \frac{1}{2} + \frac{b(\alpha+1)}{2} \,,$$

that is, $b\alpha \geq 1$ for every $b > 2$. Hence $\alpha \geq 1/2$. $\qquad\square$

Before going on we state the general result proved by Erdős and Fuchs. See [75, 130].

---

[2] If the numbers α_j and β_j are ≥ 0, then $\sum_{j=1}^{+\infty} \alpha_j \beta_j \leq \left\{\sum_{j=1}^{+\infty} \alpha_j^p\right\}^{1/p} \left\{\sum_{j=1}^{+\infty} \beta_j^q\right\}^{1/q}$, where $1 < p, q < +\infty$ and $\frac{1}{p} + \frac{1}{q} = 1$.

Theorem 8.19 (Erdős and Fuchs) *Let A be a set of non-negative integers and, for every non-negative integer n, let $\gamma(n)$ be the number of pairs $(a_1, a_2) \in A \times A$ which solve the equation $n = a_1 + a_2$. Let us assume the existence of positive numbers L, α, c such that*

$$\left| \sum_{h=0}^{n} (\gamma(h) - L) \right| \leq cn^{\alpha} .$$

Then $\alpha \geq 1/4$.

The following result was proved by Kendall in 1948 [106] (see also [35]). Kendall's result deals with the 'shifted circle problem'. It shows that the conjecture $\theta = 1/2$ is true in this easier problem. Apparently, Kendall was the first to realize that certain lattice point problems can be studied using Fourier analysis in several variables.

Theorem 8.20 (Kendall) *For every $t \in \mathbb{R}^2$ let*

$$D_R(t) := \text{card}\left(B(t, R) \cap \mathbb{Z}^2\right) - \pi R^2 = -\pi R^2 + \sum_{k \in \mathbb{Z}^2} \chi_{B(t,R)}(k) ,$$

where $B(t, R)$ is the disc with centre t and radius R. Then there exists $c > 0$ such that, for every $R \geq 1$, we have

$$\|D_R\|_{L^2(\mathbb{T}^2)} \leq c\, R^{1/2} .$$

A comparison between Kendall's result and (8.23) shows that the discs centred at the origin have a discrepancy larger than the average.

Proof The integration is over $\mathbb{T}^2$ since the function $D_R(t)$ is periodic. We compute the Fourier coefficients of $D_R(t)$. For $m = 0$ we have

$$\widehat{D}_R(0) = \int_{[0,1)^2} \left(-\pi R^2 + \sum_{k \in \mathbb{Z}^2} \chi_{B(t,R)}(k) \right) dt = -\pi R^2 + \sum_{k \in \mathbb{Z}^2} \int_{[0,1)^2} \chi_{B(t,R)}(k)\, dt$$

$$= -\pi R^2 + \sum_{k \in \mathbb{Z}^2} \int_{[0,1)^2} \chi_{B(0,R)}(k - t)\, dt = -\pi R^2 + \int_{\mathbb{R}^2} \chi_{B(0,R)}(t)dt = 0 .$$

For $0 \neq m \in \mathbb{Z}^2$ we have

$$\widehat{D}_R(m) = \int_{[0,1)^2} D_R(t)\, e^{-2\pi i m \cdot t}\, dt = \int_{[0,1)^2} \left(-\pi R^2 + \sum_{k \in \mathbb{Z}^2} \chi_{B(t,R)}(k) \right) e^{-2\pi i m \cdot t}\, dt$$

$$= \sum_{k \in \mathbb{Z}^2} \int_{[0,1)^2} \chi_{B(t,R)}(k)\, e^{-2\pi i m \cdot t}\, dt = \sum_{k \in \mathbb{Z}^2} \int_{[0,1)^2} \chi_{B(0,R)}(k - t)\, e^{-2\pi i m \cdot t}\, dt$$

$$= \int_{\mathbb{R}^2} \chi_{B(0,R)}(t)\, e^{-2\pi i m \cdot t}\, dt = \widehat{\chi}_R(m) = R^2 \widehat{\chi}_1(Rm) .$$

By Lemma 8.14 there exists $c > 0$ such that, for every $m \neq 0$,

$$\left|\widehat{D_R}(m)\right| \leq cR^{1/2}\frac{1}{|m|^{3/2}} .$$

Then Parseval's identity (4.23) implies

$$\|D_R\|^2_{L^2(\mathbb{T}^2)} = \sum_{m \in \mathbb{Z}^2} \left|\widehat{D_R}(m)\right|^2 \leq c_1 R \sum_{0 \neq m \in \mathbb{Z}^2} |m|^{-3} \leq c_2 R \int_1^{+\infty} r^{-2}\, dr = c_2 R .$$

$\square$

In the last chapter we shall prove the lower bound $\|D_R\|_{L^2(\mathbb{T}^2)} \geq c\, R^{1/2}$. Moreover we shall extend Theorem 8.20 to d variables and see that the corresponding lower bound may change significantly with d. For other mean estimates related to the circle problem see, for example, [60, 99].

8.4 Integer points in convex bodies

In this section we shall replace the disc with a square or, more generally, with a planar convex body. In the case of a square Q we observe that if the sides are parallel to the axes (or have rational slopes), then the function

$$R \mapsto R^2 - \mathrm{card}\left(RQ \cap \mathbb{Z}^2\right)$$

changes sign infinitely many times and we have

$$cR \leq \left|R^2 - \mathrm{card}\left(RQ \cap \mathbb{Z}^2\right)\right| \leq c_1 R$$

for infinitely many values of R. See the figure below, where the two squares have almost the same area, while the largest one contains $\approx R$ integer points more than the smaller one.

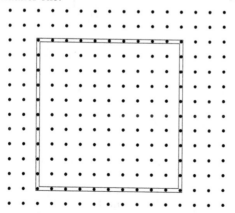

Averaging over the rotations we obtain a different estimate [28, 144, 169].

Theorem 8.21 *Let $P \subset \mathbb{R}^2$ be a polygon and let $SO(2)$ be the rotation group in $\mathbb{R}^2$. For every*

$$\sigma_\theta = \begin{bmatrix} \cos\theta & \sin\theta \\ -\sin\theta & \cos\theta \end{bmatrix} \in SO(2)$$

let $\sigma_\theta(P)$ be the polygon rotated by the angle θ. Then there exists a positive constant c such that, for every $R \geq 2$,

$$\int_0^{2\pi} \left| \text{card}\left(R\sigma_\theta(P) \cap \mathbb{Z}^2\right) - R^2 |P| \right| d\theta \leq c \log^2 R .$$

We need a lemma on the *average decay* of the Fourier transform of the characteristic function χ_P of P.

Lemma 8.22 *Let $\chi_P(t)$ be the characteristic function of a polygon P. Let $1 \leq p \leq \infty$, then there exist positive constants c and c_p such that, for every $\rho \geq 2$,*

$$\left\| \widehat{\chi_P}(\rho \cdot) \right\|_{L^p([0,2\pi))} \leq \begin{cases} c\rho^{-2} \log\rho & \text{if } p = 1, \\ c_p \rho^{-1-1/p} & \text{if } 1 < p \leq \infty, \end{cases} \tag{8.28}$$

where $\widehat{\chi_P}(\rho\Theta) = \widehat{\chi_P}(\rho\cos\theta, \rho\sin\theta)$ is written in polar coordinates.

Proof We may subdivide P as a disjoint (up to sets of measure zero) union of a finite number of convex polygons. That is, we may assume that P is convex and therefore apply Theorem 8.13. Then the case $p = \infty$ follows from (8.10). For the case $p < \infty$ we look at the figure below:

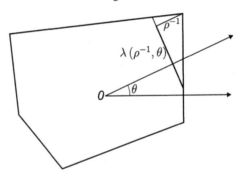

which shows that the chord at distance ρ^{-1} from the boundary (see (8.8)) has length

$$\left| \lambda\left(\rho^{-1}, \theta\right) \right| \leq c \, \min\left(1, \rho^{-1}\theta^{-1}\right) .$$

Then

$$\left\{\int_0^{2\pi} \left|\widehat{\chi}_P\left(\rho\Theta\right)\right|^p \, d\theta\right\}^{1/p} \le \frac{c_1}{\rho}\left\{\int_0^{1/\rho} d\theta\right\}^{1/p} + \frac{c_1}{\rho}\left\{\int_{1/\rho}^{c_2}\left(\frac{1}{\rho\theta}\right)^p d\theta\right\}^{1/p}$$

$$\le \begin{cases} c_3\rho^{-2}\log\rho & \text{if } p = 1, \\ c_p\rho^{-1-1/p} & \text{if } p > 1, \end{cases}$$

where c_p depends on p. $\qquad\square$

A different proof of the previous result will be given implicitly in the proof of Lemma 10.6. The following lemma was pointed out to us by Colzani.

Lemma 8.23 *Let $C \subset \mathbb{R}^d$ be a closed convex body. Let A^o denote the interior of a set $A \subseteq \mathbb{R}^d$ and let $\partial(A)$ be the boundary of A. Then for large R and small ε we have, for every $q \in \partial(RC)$,*

$$B(q, \varepsilon) \subseteq (R + \varepsilon)C \setminus ((R - \varepsilon)C)^o \,,$$

where $B(q, \varepsilon) = \left\{t \in \mathbb{R}^d : |t - q| \le \varepsilon\right\}$.

Proof Since C has positive measure, there is an open ball contained in C. We may assume that $B(0, 1) \subset C$. By convexity we have

$$\frac{R}{R + \varepsilon}C + \frac{\varepsilon}{R + \varepsilon}C \subseteq C \,.$$

Hence

$$(R + \varepsilon)C \supseteq RC + \varepsilon C \supseteq RC + B(0, \varepsilon) \,, \qquad (8.29)$$

and therefore $B(q, \varepsilon) \subseteq (R + \varepsilon)C$ for every $q \in \partial(RC)$. We now replace $R + \varepsilon$ with R and apply (8.29) to C^o. Then

$$((R - \varepsilon)C)^o + B(0, \varepsilon) \subseteq (RC)^o \,.$$

If $y \in B(q, \varepsilon) \cap ((R - \varepsilon)C)^o$, then $q \in (RC)^o$. Hence $q \notin \partial(RC)$. $\qquad\square$

Proof of Theorem 8.21 We may assume that P is convex. Let χ_P be the characteristic function of P. We follow the argument in the proof of Theorem 8.17. Let

$$\varphi_\varepsilon(t) := \pi^{-1}\varepsilon^{-2}\chi_\varepsilon(t) \,,$$

$$\widetilde{\chi}_R^{(\varepsilon)}(t) := (\varphi_\varepsilon * \chi_{RP})(t) \,, \quad \widetilde{N}_R^{(\varepsilon,\theta)} := \sum_{m\in\mathbb{Z}^2} \widetilde{\chi}_R^{(\varepsilon)}(\sigma_\theta(m)) \,.$$

We observe that

$$\widehat{\varphi}_\varepsilon(\sigma_\theta(m)) = \pi^{-1}\varepsilon^{-2}\int_{|t|\le\varepsilon} e^{-2\pi i \sigma_\theta(m)\cdot t} \, dt = \pi^{-1}\varepsilon^{-2}\int_{|t|\le\varepsilon} e^{-2\pi i m \cdot \sigma_\theta^{-1}(t)} \, dt$$

$$= \pi^{-1} \varepsilon^{-2} \int_{|t| \le \varepsilon} e^{-2\pi i m \cdot t} \, dt = \widehat{\varphi_\varepsilon}(m) = \pi^{-1} \widehat{\chi_1}(\varepsilon m) \, .$$

Moreover,

$$(\chi_{RP} \left(\sigma_\theta(\cdot) \right))^\wedge (m) = \int_{\mathbb{R}^2} \chi_{RP} \left(\sigma_\theta(t) \right) e^{-2\pi i m \cdot t} \, dt$$

$$= \int_{\mathbb{R}^2} \chi_{RP} (u) \, e^{-2\pi i m \cdot \sigma_\theta^{-1}(u)} \, du = \int_{\mathbb{R}^2} \chi_{RP} (u) \, e^{-2\pi i \sigma_\theta(m) \cdot u} \, du$$

$$= R^2 \int_{\mathbb{R}^2} \chi_P (s) \, e^{-2\pi i R \sigma_\theta(m) \cdot s} \, ds = R^2 \, \widehat{\chi_P}(R \sigma_\theta(m)) \, .$$

Then Lemma 8.14 and (8.10) give, for $|\xi| \ge 1$,

$$\left| \left(\widetilde{\chi}_R^{(\varepsilon)}(\sigma_\theta(\cdot)) \right)^\wedge (\xi) \right| = \left| \widetilde{\chi}_R^{(\varepsilon)} \left(\sigma_\theta^{-1}(\xi) \right) \right| = \left| \widehat{\varphi_\varepsilon} \left(\sigma_\theta^{-1}(\xi) \right) \widehat{\chi_{RP}} \left(\sigma_\theta^{-1}(\xi) \right) \right|$$

$$= R^2 \left| \widehat{\chi_1}(\xi) \, \widehat{\chi_P} \left(R \sigma_\theta^{-1}(\xi) \right) \right| \le cR \, |\xi|^{-5/2} \, \varepsilon^{-3/2} \, ,$$

where c is independent of θ. Hence we can apply the Poisson summation formula and obtain

$$\widetilde{N}_R^{(\varepsilon,\theta)} = \sum_{m \in \mathbb{Z}^2} \left(\widetilde{\chi}_R^{(\varepsilon)}(\sigma_\theta(\cdot)) \right)^\wedge (m) = \sum_{m \in \mathbb{Z}^2} \widehat{\varphi_\varepsilon}(m) \, (\chi_{RP}(\sigma_\theta(\cdot)))^\wedge (m) \qquad (8.30)$$

$$= R^2 \sum_{m \in \mathbb{Z}^2} \widehat{\varphi_\varepsilon}(m) \, \widehat{\chi_P}(R\sigma_\theta(m)) = R^2 \, |P| + R^2 \sum_{m \ne 0} \widehat{\varphi_\varepsilon}(m) \, \widehat{\chi_P}(R\sigma_\theta(m)) \, .$$

By Lemma 8.23 we have

$$\widetilde{\chi}_R^{(\varepsilon)}(t) = (\varphi_\varepsilon * \chi_{RP})(t) = \int_{RP} \varphi_\varepsilon(t-s) \, ds = \begin{cases} 1 & \text{if } t \in (R-\varepsilon)P, \\ 0 & \text{if } t \notin (R+\varepsilon)P. \end{cases}$$

Then

$$\widetilde{N}_{R-\varepsilon}^{(\varepsilon,\theta)} \le \text{card} \left(R\sigma_\theta(P) \cap \mathbb{Z}^2 \right) = \sum_{m \in \mathbb{Z}^2} \chi_{RP}(\sigma_\theta(m)) \le \widetilde{N}_{R+\varepsilon}^{(\varepsilon,\theta)} \qquad (8.31)$$

and therefore (8.28), (8.30) and (8.31) imply

$$c \int_0^{2\pi} \left| \text{card} \left(R\sigma_\theta(P) \cap \mathbb{Z}^2 \right) - R^2 \, |P| \right| \, d\theta$$

$$\le 2R\varepsilon + \varepsilon^2 + \max_{\pm} \left((R \pm \varepsilon)^2 \sum_{m \ne 0} \left| \widehat{\varphi_\varepsilon}(m) \right| \int_0^{2\pi} \left| \widehat{\chi_P} \left((R \pm \varepsilon) \, |m| \, \sigma_\theta \left(\frac{m}{|m|} \right) \right) \right| \, d\theta \right)$$

$$\le 2R\varepsilon + c_1 R^2 \sum_{m \ne 0} \left| \widehat{\varphi_\varepsilon}(m) \right| (R \, |m|)^{-2} \log (R \, |m|) \, .$$

Now we choose $\varepsilon = 1/R$. Then (8.19) and Lemma 8.14 give

$$R^2 \sum_{m \ne 0} \left| \widehat{\varphi_{1/R}}(m) \right| (R \, |m|)^{-2} \log (R \, |m|)$$

$$= \pi^{-1} \sum_{m \neq 0} \left| \widehat{\chi}_1(R^{-1}m) \right| |m|^{-2} \log(R\,|m|)$$

$$\leq c_2 \sum_{m \neq 0} \frac{1}{(1 + |m|\,/R)^{3/2}} |m|^{-2} \log(R\,|m|)$$

$$\leq c_3 \int_1^\infty \frac{1}{x\,(1 + x/R)^{3/2}} \log(Rx)\, dx$$

$$\leq c_4 \int_1^R x^{-1} \log(R^2)\, dx + c_4 R^{3/2} \int_R^\infty x^{-5/2} \log\left(x^2\right) dx \leq c_8 \log^2 R \;.$$

$\square$

The following result, proved by Davenport in 1958 [67], shows that suitable rotations make the discrepancy of a square at most logarithmic.

Theorem 8.24 (Davenport) *Let α be an irrational algebraic number of degree 2 and let Q be a unit square with a side parallel to the vector $(1, a)$. Then there exists $c > 0$ such that*

$$\int_{\mathbb{T}^2} \left| \mathrm{card}\left(R(Q + t) \cap \mathbb{Z}^2 \right) - R^2 \right|^2 dt \leq c \log R \;.$$

Proof For simplicity, let us assume that the unit square Q has sides parallel to the unit vectors $\sigma = \left(-1/\sqrt{3},\ \sqrt{2}/\sqrt{3} \right)$ and $\sigma^\perp = \left(\sqrt{2}/\sqrt{3},\ 1/\sqrt{3} \right)$ respectively. Arguing as in the proof of Theorem 8.20, we write

$$\int_{\mathbb{T}^2} \left| \mathrm{card}\left(R(Q + t) \cap \mathbb{Z}^2 \right) - R^2 \right|^2 dt$$

$$= R^4 \sum_{m \neq 0} \left| \widehat{\chi}_Q(Rm) \right|^2$$

$$\leq R^4 \left\{ \sum_{\substack{m \neq 0,\ |m \cdot \sigma| \leq R^{-1}}} + \sum_{R^{-1} < |m \cdot \sigma| \leq \frac{1}{2\sqrt{3}}} \right.$$

$$+ \sum_{\substack{m \neq 0,\ |m \cdot \sigma^\perp| \leq R^{-1}}} + \sum_{R^{-1} < |m \cdot \sigma^\perp| \leq \frac{1}{2\sqrt{3}}} + \left. \sum_{\frac{1}{2\sqrt{3}} < |m \cdot \sigma|,\ \frac{1}{2\sqrt{3}} < |m \cdot \sigma^\perp|} \right\}$$

$$:= R^4 \left\{ A + B + C + D + E \right\} \;.$$

By symmetry it is enough to estimate A, B and E. We start with the latter. By (8.11) we have

$$\widehat{\chi}_Q(Rm) = \frac{\sin\left(\pi R\sigma \cdot m \right)}{\pi R\sigma \cdot m} \frac{\sin(\pi R\sigma^\perp \cdot m)}{\pi R\sigma^\perp \cdot m} \;, \tag{8.32}$$

and therefore

$$R^4 E \leq c_1 \sum_{\frac{1}{2\sqrt{3}} < |m \cdot \sigma|, \ \frac{1}{2\sqrt{3}} < |m \cdot \sigma^\perp|} \frac{1}{|\sigma \cdot m|^2} \frac{1}{|\sigma^\perp \cdot m|^2}$$

$$\leq c_2 \int_1^{+\infty} \int_1^{+\infty} \frac{dx dy}{x^2 y^2} \leq c_3 .$$

We bound A. By (8.32) we have

$$R^4 A \leq R^4 \sum_{m \neq 0, \ |m \cdot \sigma| \leq R^{-1}} \left| \frac{\sin(\pi R \sigma^\perp \cdot m)}{\pi R \sigma^\perp \cdot m} \right|^2 \leq c_4 R^2 \sum_{m \neq 0, \ |m \cdot \sigma| \leq R^{-1}} \frac{1}{|m|^2} .$$

The set

$$\left\{ m \in \mathbb{Z}^2 : |m \cdot \sigma| \leq R^{-1} \right\} = \left\{ m \in \mathbb{Z}^2 : \left| m_1 - m_2 \sqrt{2} \right| \leq \sqrt{3} R^{-1} \right\}$$

consists of integer points inside a strip of width $2R^{-1}$ about the line $\mathbb{R}\sigma^\perp$. Since we may assume that $R > 2\sqrt{3}$, the condition $\left| m_1 - m_2 \sqrt{2} \right| \leq \sqrt{3} R^{-1}$ implies $\left| m_1 - m_2 \sqrt{2} \right| < 1/2$. Then for every m_2 there exists at most one m_1 such that $\left| m_1 - m_2 \sqrt{2} \right| \leq \sqrt{3} R^{-1}$. We may assume that $m_2 > 0$. For simplicity we replace R with $R \sqrt{3}$. Then

$$R^4 A \leq c_4 R^2 \sum_{m \neq 0, \ \left| m_1 - m_2 \sqrt{2} \right| \leq \sqrt{3} R^{-1}} \frac{1}{|m|^2} \leq c_5 R^2 \sum_{1 \leq k < +\infty, \ \left\| k \sqrt{2} \right\| \leq R^{-1}} \frac{1}{k^2} , \quad (8.33)$$

where $\|\alpha\|$ is the distance of α from the integers. By Theorem 5.6 there exists $H > 0$ such that $\left\| k \sqrt{2} \right\| \geq H/k$ for every positive integer k. Then we must have $R^{-1} \geq H/k$ in (8.33). Hence the last term in (8.33) is bounded by

$$R^2 \sum_{HR \leq k < +\infty, \ \left\| k \sqrt{2} \right\| \leq R^{-1}} \frac{1}{k^2} \leq c_6 R^2 \sum_{\log R \leq q < +\infty} \left(\sum_{2^{q-1} \leq k < 2^q, \ \left\| k \sqrt{2} \right\| \leq R^{-1}} 2^{-2q} \right) .$$

As in the proof of Theorem 7.4, we observe that if $2^{q-1} \leq k < 2^q$ then, for every positive integer $s < 2^q/H$, each interval of the form $\left[\frac{sH}{2^q}, \frac{(s+1)H}{2^q} \right)$ contains at most two numbers of the form $\left\| k \sqrt{2} \right\|$. Assume that there are three of them. Then, say, we have two of them of the form $\left\{ k_1 \sqrt{2} \right\}$ and $\left\{ k_2 \sqrt{2} \right\}$, with $2^{q-1} \leq k_1 < k_2 < 2^q$. Assuming q large, Theorem 5.6 implies

$$\frac{H}{2^q} \geq \left| \left\{ k_2 \sqrt{2} \right\} - \left\{ k_1 \sqrt{2} \right\} \right| = \left\| (k_2 - k_1) \sqrt{2} \right\| \geq \frac{H}{k_2 - k_1} > \frac{H}{2^q} .$$

Since the inequalities

$$\frac{sH}{2^q} \leq \left\| k \sqrt{2} \right\| < \frac{(s+1)H}{2^q} , \quad \left\| k \sqrt{2} \right\| \leq R^{-1}$$

imply $\frac{sH}{2^q} \leq R^{-1}$, the last term in (8.4) is not larger than

$$c_7 R^2 \sum_{\log R \leq q < +\infty} 2^{-2q} \sum_{1 \leq s \leq \frac{2^q}{HR}} 1 = c_8 R \sum_{\log R \leq q < +\infty} 2^{-q} \leq c_9 \ .$$

We now consider B. Since $\left\| k\sqrt{2} \right\| \geq H/k \geq R^{-1}$ we have

$$R^4 B \leq c_{10} R^4 \sum_{R^{-1} < \left| m_1 - m_2 \sqrt{2} \right| \leq 1/2} \left| \frac{\sin(\pi R\sigma \cdot m)}{\pi R\sigma \cdot m} \frac{\sin(\pi R\sigma^\perp \cdot m)}{\pi R\sigma^\perp \cdot m} \right|^2$$

$$\leq c_{10} \sum_{R^{-1} < \left| m_1 - m_2 \sqrt{2} \right| \leq 1/2} \frac{1}{\left| \sigma \cdot m \right|^2} \frac{1}{\left| m \right|^2} \leq c_{11} \sum_{1 \leq k < +\infty, \ R^{-1} < \left\| k\sqrt{2} \right\|} \frac{1}{\left\| k\sqrt{2} \right\|^2} \frac{1}{k^2}$$

$$= c_{11} \sum_{1 \leq k \leq HR, \ R^{-1} < \left\| k\sqrt{2} \right\|} \frac{1}{\left\| k\sqrt{2} \right\|^2} \frac{1}{k^2} + c_{11} \sum_{HR < k < +\infty, \ R^{-1} < \left\| k\sqrt{2} \right\|} \frac{1}{\left\| k\sqrt{2} \right\|^2} \frac{1}{k^2} \ .$$

Then, arguing as before,

$$\sum_{1 \leq k \leq HR} \frac{1}{\left\| k\sqrt{2} \right\|^2} \frac{1}{k^2} \leq c_{12} \sum_{1 \leq q \leq \log R} \sum_{2^{q-1} \leq k < 2^q - 1} \frac{2^{-2q}}{\left\| k\sqrt{2} \right\|^2}$$

$$\leq c_{13} \sum_{1 \leq q \leq \log R} 2^{-2q} \sum_{1 \leq s \leq 2^q/H} \left(\frac{2^q}{sH} \right)^2 \leq c_{14} \log R \ .$$

Finally,

$$\sum_{HR < k < +\infty, \ R^{-1} < \left\| k\sqrt{2} \right\|} \frac{1}{\left\| k\sqrt{2} \right\|^2} \frac{1}{k^2}$$

$$\leq c_{15} \sum_{\log R < q < +\infty} 2^{-2q} \sum_{2^{q-1} \leq k < 2^q - 1, \ R^{-1} < \left\| k\sqrt{2} \right\|} \frac{1}{\left\| k\sqrt{2} \right\|^2}$$

$$\leq c_{16} \sum_{\log R < q < +\infty} 2^{-2q} \sum_{\frac{2^q}{RH} \leq s \leq \frac{2^q}{H}} \left(\frac{2^q}{sH} \right)^2 \leq c_{17} \sum_{\log R < q < +\infty} \sum_{\frac{2^q}{RH} \leq s \leq \frac{2^q}{H}} \frac{1}{s^2}$$

$$\leq c_{18} \sum_{\log R < q < +\infty} \int_{\frac{2^q}{RH}}^{+\infty} \frac{1}{t^2} dt \leq c_{19} R \sum_{\log R < q < +\infty} 2^{-q} \leq c_{20} \ .$$

□

We now go back to (8.28) and show that the estimate for $p = 2$ is a particular case of a general result, proved by Podkorytov [142] (see also [27, 30, 32, 33]).

First we show how to construct a partition of unity on a closed smooth curve, say for simplicity the unit circle $\mathbb{T}$. We choose $n \geq 2$ points $x_1, x_2, \ldots, x_n \in \mathbb{T}$

and n small positive numbers $r_1, r_2, \ldots, r_n$ such that $x_j \notin (x_k - r_k, x_k + r_k)$ when $j \neq k$, and the union of the open intervals $(x_j - r_j, x_j + r_j)$ covers $\mathbb{T}$. For every $u \in \mathbb{R}$ let

$$\lambda(u) := \begin{cases} e^{-1/(1-u^2)} & \text{if } -1 < u < 1, \\ 0 & \text{if } |u| \geq 1. \end{cases}$$

Then $\lambda(u)$ is smooth on $\mathbb{R}$. Indeed, we have $e^{-\frac{1}{1-u^2}} = e^{-\frac{1}{2(1-u)}} e^{-\frac{1}{2(1+u)}}$ and it is easy to see that the function

$$g(x) := \begin{cases} e^{-1/x} & \text{if } x > 0, \\ 0 & \text{if } x \leq 0 \end{cases}$$

belongs to $C^\infty(\mathbb{R})$. For every $x \in \mathbb{T}$ and $j = 1, \ldots, n$, let

$$\varphi_j(x) := \lambda\left(\frac{|x - x_j|}{r_j}\right).$$

Then

$$\psi_j(x) := \frac{\varphi_j(x)}{\varphi_1(x) + \varphi_2(x) + \ldots + \varphi_n(x)}$$

belongs to $C^\infty(\mathbb{T})$ and satisfies, for every $x \in \mathbb{T}$,

$$\psi_1(x) + \psi_2(x) + \ldots + \psi_n(x) = 1 ,$$

while $\psi_j(x) = 1$ in a neighbourhood of x_j. We say that $\{\psi_j(x)\}_{j=1}^n$ is a *partition of unity* on $\mathbb{T}$.

Theorem 8.25 (Podkorytov) *Let $C \subset \mathbb{R}^2$ be a convex body. For $0 \leq \theta < 2\pi$ we write $\Theta := (\cos\theta, \sin\theta)$. Then there is a positive constant c such that, for every $\rho \geq 1$,*

$$\int_0^{2\pi} |\widehat{\chi_C}(\rho\Theta)|^2 \, d\theta \leq c\,\rho^{-3} .$$

Proof We may assume that ∂C is smooth as long as the constant c does not depend on the regularity of ∂C. For $0 \neq \xi \in \mathbb{R}^2$ let

$$\omega(t) = \frac{e^{-2\pi i \xi \cdot t}}{-2\pi i |\xi|^2} \xi .$$

Then

$$\operatorname{div}\omega(t) = \frac{\partial}{\partial t_1}\left(\frac{e^{-2\pi i(\xi_1 t_1 + \xi_2 t_2)}}{-2\pi i |\xi|^2}\xi_1\right) + \frac{\partial}{\partial t_2}\left(\frac{e^{-2\pi i(\xi_1 t_1 + \xi_2 t_2)}}{-2\pi i |\xi|^2}\xi_2\right) = e^{-2\pi i \xi \cdot t}$$

and the divergence theorem implies

$$\widehat{\chi_C}(\rho\Theta) = \int_C e^{-2\pi i \rho\Theta \cdot t}\, dt = -\frac{1}{2\pi i \rho} \int_{\partial C} e^{-2\pi i \rho\Theta \cdot t} \, (\nu(t) \cdot \Theta)\, ds(t) ,$$

where $v(t)$ is the outward unit normal to ∂C at t. Let η_j be a finite, smooth, non-negative partition of unity on ∂C. Assume that each function η_j has support in $\Omega_j \subset \partial C$. We then have to prove that

$$\int_0^{2\pi} \left| \int_\Omega e^{-2\pi i \rho \Theta \cdot t} \, (v(t) \cdot \Theta) \, \eta(t) \, ds(t) \right|^2 \, d\theta \leq c \, \rho^{-1} \, ,$$

where we have written η and Ω in place of η_j and Ω_j, respectively. We may assume that Ω is as in the figure below:

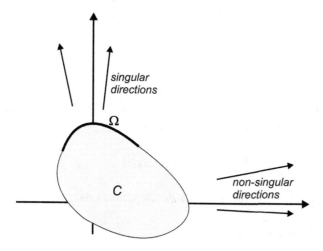

We split $[0, 2\pi)$ into directions essentially orthogonal to Ω ('singular') and directions essentially non-orthogonal to Ω ('non-singular'). More precisely, let $0 < c_1 < c_2 < 1$ and let $\gamma : \mathbb{R} \to [0, 1]$ be a smooth function with period 2 such that $\gamma(x) = 1$ if $|x| < c_1$ and $\gamma(x) = 0$ if $c_2 < |x| \leq 1$. Then we write

$$\int_0^{2\pi} \left| \int_\Omega e^{-2\pi i \rho \Theta \cdot t} \, (v(t) \cdot \Theta) \, \eta(t) \, ds(t) \right|^2 \, d\theta$$

$$= \int_0^{2\pi} \left| \int_\Omega e^{-2\pi i \rho \Theta \cdot t} \, (v(t) \cdot \Theta) \, \eta(t) \, ds(t) \right|^2 \gamma(\cos\theta) \, d\theta$$

$$+ \int_0^{2\pi} \left| \int_\Omega e^{-2\pi i \rho \Theta \cdot t} \, (v(t) \cdot \Theta) \, \eta(t) \, ds(t) \right|^2 (1 - \gamma(\cos\theta)) \, d\theta$$

$$:= \mathfrak{S} + \mathfrak{N}\mathfrak{S} \, .$$

We start with $\mathfrak{N}\mathfrak{S}$. In this case the function $\theta \mapsto \Theta \cdot t$ is not stationary. Since C is convex, then Lemma 8.11 and a change of variables allow us to write

$$\int_\Omega e^{-2\pi i \rho \Theta \cdot t} \, (v(t) \cdot \Theta) \, \eta(t) \, ds(t)$$

as

$$\int_a^b e^{-2\pi i \rho \varphi(x)} \, \psi(x) \, dx \,,$$

where $\varphi'(x)$ is smooth and away from 0 in (a, b), while $\psi(x)$ is smooth and supported in (a, b). Then we can integrate by parts to obtain

$$\int_a^b e^{-2\pi i \rho \varphi(x)} \, \psi(x) \, dx = \int_a^b \frac{1}{-2\pi i \rho \varphi'(x)} \frac{d}{dx} \left(e^{-2\pi i \rho \varphi(x)} \right) \psi(x) \, dx \qquad (8.34)$$

$$= \int_a^b e^{-2\pi i \rho \varphi(x)} \frac{d}{dx} \left(\frac{\psi(x)}{2\pi i \rho \varphi'(x)} \right) dx \,.$$

Hence

$$\Re \mathfrak{S} \leq c \, \rho^{-1} \,.$$

The previous argument does not work for $\mathfrak{S}$, because $\Theta \cdot t$ can be stationary. Instead, we use the fact that we are dealing with an L^2 norm:

$$\mathfrak{S} = \int_0^{2\pi} \int_\Omega e^{-2\pi i \rho \Theta \cdot t} \, (v(t) \cdot \Theta) \, \eta(t) \, ds(t)$$

$$\times \int_\Omega e^{2\pi i \rho \Theta \cdot u} \, (v(u) \cdot \Theta) \, \eta(u) \, ds(u) \, \gamma(\cos\theta) \, d\theta$$

$$= \int_\Omega \eta(t) \int_\Omega \eta(u)$$

$$\times \int_0^{2\pi} e^{2\pi i \rho \Theta \cdot (u-t)} \, (v(t) \cdot \Theta) \, (v(u) \cdot \Theta) \gamma(\cos\theta) \, d\theta \, ds(t) \, ds(u) \,.$$

The function

$$\tau(\theta) := (v(t) \cdot \Theta) \, (v(u) \cdot \Theta) \, \gamma(\cos\theta)$$

is smooth with small support (contained, say, in $(-\delta + \pi/2, \delta + \pi/2)$). Observe that the function $\theta \mapsto \Theta \cdot (u - t)$ is not stationary on this interval. If we argue as before we need to study

$$\int_{-\delta+\pi/2}^{\delta+\pi/2} e^{2\pi i \rho \Theta \cdot (u-t)} \, \tau(\theta) \, d\theta = \int_{-\delta+\pi/2}^{\delta+\pi/2} e^{2\pi i \rho \Theta \cdot (u-t)} \frac{d}{d\theta} \left(\frac{\tau(\theta)}{2\pi i \rho \Theta' \cdot (u-t)} \right) d\theta$$

$$= \int_{-\delta+\pi/2}^{\delta+\pi/2} e^{2\pi i \rho \Theta \cdot (u-t)} \frac{d}{d\theta} \left(\frac{1}{2\pi i \rho \Theta' \cdot (u-t)} \frac{d}{d\theta} \left(\frac{\tau(\theta)}{2\pi i \rho \Theta' \cdot (u-t)} \right) \right) d\theta \,.$$

Then

$$\left| \int_{-\delta+\pi/2}^{\delta+\pi/2} e^{2\pi i \rho \Theta \cdot (u-t)} \, \tau(\theta) \, d\theta \right| \leq c_1 \frac{1}{1 + \rho^2 |u - t|^2} \,.$$

Finally,

$$\mathfrak{S} \leq c_2 \int_\Omega \int_\Omega \frac{1}{1 + \rho^2 |u - t|^2} \, ds(t) \, ds(u) \tag{8.35}$$

$$\leq c_2 \int_{|u-t| < \rho^{-1}} ds(t) \, ds(u) + c_2 \, \rho^{-2} \int_{\rho^{-1} \leq |u-t| \leq c_3} \frac{1}{|u - t|^2} \, ds(t) \, ds(u)$$

$$\leq c_4 \rho^{-1} + c_4 \rho^{-2} \int_{\rho^{-1}}^{c_5} \frac{1}{x^2} \, dx \leq c_6 \, \rho^{-1} .$$

This completes the proof. $\qquad\square$

Now we can extend Kendall's theorem to the case of an arbitrary planar convex body.

Theorem 8.26 *Let $C \subset \mathbb{R}^2$ be a convex body. Then there exists $c > 0$ such that, for every $R \geq 1$,*

$$\left\{ \int_{SO(2)} \int_{\mathbb{T}^2} \left| \mathrm{card}\left(\mathbb{Z}^2 \cap (R\sigma_\theta(C) + t)\right) - R^2 |C| \right|^2 \, dt \, d\sigma \right\}^{1/2} \leq cR^{1/2} .$$

Proof We argue as in Theorem 8.20. Let

$$D_{R,\theta}(t) := \mathrm{card}\left(\mathbb{Z}^2 \cap (R\sigma_\theta(C) + t)\right) - R^2 |C| \tag{8.36}$$

$$= -R^2 |C| + \sum_{k \in \mathbb{Z}^2} \chi_{\sigma_\theta(RC)}(k - t) .$$

We compute the Fourier coefficients of the periodic function $D_{R,\theta}(t)$:

$$\widehat{D}_{R,\theta}(0) = \int_{\mathbb{T}^2} \left(\mathrm{card}\left(\mathbb{Z}^2 \cap (R\sigma_\theta(C) + t)\right) - R^2 |C| \right) dt$$

$$= -R^2 |C| + \sum_{k \in \mathbb{Z}^d} \int_{\mathbb{T}^2} \chi_{\sigma_\theta(RC)}(k - t) \, dt = -R^2 |C| + \int_{\mathbb{R}^2} \chi_{\sigma_\theta(RC)}(t) \, dt = 0 ,$$

while for $0 \neq m \in \mathbb{Z}^2$ we have

$$\widehat{D}_{R,\theta}(m) = \int_{\mathbb{T}^2} \left(\mathrm{card}\left(\mathbb{Z}^2 \cap (R\sigma_\theta(C) + t)\right) - R^2 |C| \right) e^{-2\pi i m \cdot t} \, dt$$

$$= \sum_{k \in \mathbb{Z}^d} \int_{\mathbb{T}^2} \chi_{\sigma_\theta(RC)}(k - t) \, e^{-2\pi i m \cdot t} \, dt = \int_{\mathbb{R}^2} \chi_{\sigma_\theta(RC)}(t) \, e^{-2\pi i m \cdot t} \, dt$$

$$= \widehat{\chi}_{\sigma_\theta(RC)}(m) = R^2 \, \widehat{\chi}_C \left(R\sigma_\theta^{-1}(m) \right) .$$

Then $D_{R,\theta}(t)$ has Fourier series

$$R^2 \sum_{0 \neq m \in \mathbb{Z}^2} \widehat{\chi}_C \left(R\sigma_\theta^{-1}(m) \right) e^{2\pi i m \cdot t} .$$

Note that the identity $\widehat{D}_{R,\theta}(m) = R^2 \,\widehat{\chi}_C\left(R\sigma_\theta^{-1}(m)\right)$ contains a Fourier coefficient of a periodic function on the LHS and a Fourier transform on $\mathbb{R}^2$ on the RHS. Parseval's identity and Theorem 8.25 give

$$\int_{SO(2)}\int_{\mathbb{T}^2}|D_{R,\theta}(t)|^2\,dt d\sigma = R^4\int_{SO(2)}\sum_{m\neq 0}\left|\widehat{\chi}_C\left(R\sigma_\theta^{-1}(m)\right)\right|^2\,d\sigma$$

$$= R^4\sum_{m\neq 0}\int_{SO(2)}\left|\widehat{\chi}_C\left(R\sigma_\theta^{-1}(m)\right)\right|^2\,d\sigma \leq R^4\sum_{m\neq 0}|Rm|^{-3}$$

$$\leq cR\int_{|t|\geq 1}\frac{1}{|t|^3}\,dt = c_1\,R\,.$$

$\square$

The previous two theorems hold true in several variables [30, 32].

In the case of a polygon we can fill the gap between the L^1 estimate of Theorem 8.21 and the L^2 estimate of Theorem 8.26.

Theorem 8.27   *Let $1 < p \leq +\infty$ and let P be a polygon. Then*

$$\left\{\int_{SO(2)}\int_{\mathbb{T}^2}\left|\mathrm{card}\left(\mathbb{Z}^2\cap R\sigma_\theta(P)+t\right)-R^2\,|P|\right|^p\,dt d\theta\right\}^{1/p}\leq c_p\,R^{1-1/p}\,. \quad (8.37)$$

We need a known lemma.

Lemma 8.28   *Let $a_1, a_2, \ldots, a_n$ be positive numbers. Then for every $0 < r < s$ we have*

$$\left(\sum_{j=1}^n a_j^s\right)^{1/s} < \left(\sum_{j=1}^n a_j^r\right)^{1/r}\,. \quad (8.38)$$

Proof   We have to prove

$$\sum_{j=1}^n\left(\frac{a_j}{\left(\sum_{i=1}^n a_i^r\right)^{1/r}}\right)^s < 1\,. \quad (8.39)$$

Observe that for every positive integer k we have

$$\frac{a_k}{\left(\sum_{i=1}^n (a_i)^r\right)^{1/r}} < \frac{a_k}{\left(a_k^r\right)^{1/r}} = 1\,.$$

Since $0 < r < s$ we deduce that

$$\left(\frac{a_k}{\left(\sum_{i=1}^n a_i^r\right)^{1/r}}\right)^s < \left(\frac{a_k}{\left(\sum_{i=1}^n a_i^r\right)^{1/r}}\right)^r\,.$$

Hence

$$\sum_{k=1}^{n}\left(\frac{a_k}{\left(\sum_{i=1}^{n}a_i^r\right)^{1/r}}\right)^s < \sum_{k=1}^{n}\left(\frac{a_k}{\left(\sum_{i=1}^{n}a_i^r\right)^{1/r}}\right)^r = 1 \ .$$

This proves (8.39) and (8.38). $\square$

Proof of Theorem 8.27   We start by proving the case $1 < p \le 2$. Let

$$D_R(\theta, t) = \operatorname{card}\left(\mathbb{Z}^2 \cap R\sigma_\theta(P) + t\right) - R^2 |P| \ .$$

By Parseval's identity, Corollary 6.6, Lemmas 8.28 and 8.22 we have

$$\left\{\int_{SO(2)}\int_{\mathbb{T}^2}|D_R(\theta,t)|^p \ dt d\theta\right\}^{1/p} \le \left\{\int_{SO(2)}\left\{\int_{\mathbb{T}^2}|D_R(\theta,t)|^2 \ dt\right\}^{p/2} d\theta\right\}^{1/p}$$

$$= \left\{\int_{SO(2)}\left\{\sum_{0 \ne k \in \mathbb{Z}^2}|R^2 \widehat{\chi}_{\sigma_\theta(P)}(Rk)|^2\right\}^{p/2} d\theta\right\}^{1/p}$$

$$\le \left\{\int_{SO(2)}\sum_{0 \ne k \in \mathbb{Z}^2}|R^2 \widehat{\chi}_{\sigma_\theta(P)}(Rk)|^p \ d\theta\right\}^{1/p}$$

$$= R^2 \left\{\sum_{0 \ne k \in \mathbb{Z}^2}\int_{SO(2)}\left|\widehat{\chi}_{\sigma_\theta(P)}(Rk)\right|^p \ d\theta\right\}^{1/p}$$

$$\le R^2 \left\{\sum_{0 \ne k \in \mathbb{Z}^2}|Rk|^{-p-1}\right\}^{1/p} \le R^{1-1/p}\int_1^{+\infty}s^{-p} \ ds = c_p \ R^{1-1/p} \ .$$

The case $p = +\infty$ is trivial. The case $2 < p < +\infty$ follows by an interpolation argument. Indeed, Hölder's inequality implies

$$\left\{\int_{SO(2)}\int_{\mathbb{T}^2}|D_R(\theta,t)|^p \ dt d\theta\right\}^{1/p}$$

$$= \left\{\int_{SO(2)}\int_{\mathbb{T}^2}|D_R(\theta,t)|^2 |D_R(\theta,t)|^{p-2} \ dt d\theta\right\}^{1/p}$$

$$\le \|D_R\|_{L^\infty(SO(2) \times \mathbb{T}^2)}^{(p-2)/p} \|D_R\|_{L^2(SO(2) \times \mathbb{T}^2)}^{2/p} \le c \ R^{(p-2)/p}R^{1/p} = c \ R^{1-1/p} \ .$$

$\square$

We shall see in Chapter 10 that (8.37) is the best possible.

The average over rotations can be avoided under additional geometric assumptions. The following result was proved in [59].

Theorem 8.29   *Let $C \subset \mathbb{R}^2$ be a convex body such that ∂C is smooth with everywhere positive curvature. Let*

$$D(R,t) := \operatorname{card}\left((RC - t) \cap \mathbb{Z}^2 \right) - R^2 |C| \, .$$

Then

$$\left\{ \int_{\mathbb{T}^2} |D(R,t)|^p \, dt \right\}^{1/p} \leq \begin{cases} c_p \, R^{1/2} & \text{if } p < 4, \\ c \, R^{1/2} \, \log^{3/4}(R) & \text{if } p = 4, \\ c_p \, R^{2(p-2)/(3p-4)} & \text{if } p > 4. \end{cases}$$

Exercises

1) Let $f \in L^1\left(\mathbb{R}^d\right)$ such that $f(x) \geq 0$ a.e. on $\mathbb{R}^d$. Prove the existence of $y \in \mathbb{R}^d$ such that

$$\sum_{k \in \mathbb{Z}^d} f(y+k) \geq \int_{\mathbb{R}^d} f(x) \, dx \, .$$

2) Prove that Theorem 8.13 does not extend to the case of higher variables.

3) For every $x \in \mathbb{R}$ let $f(x) = e^{-2\pi a|x|}$, with $a > 0$. Compute the Fourier transform of f and deduce the value of the sum

$$\sum_{n=1}^{+\infty} \frac{1}{n^2 + a^2} \, .$$

4) Use Theorem 5.21 to prove that for every $0 < s < 1$ we have

$$\sum_{m=0}^{+\infty} r(m) \, s^m = 1 + 4 \sum_{k=1}^{+\infty} \left(\frac{s^k}{1 - s^{4k}} - \frac{s^{3k}}{1 - s^{4k}} \right) \, .$$

5) Let P_N be a regular polygon with N sides. Prove that for no constant c independent of N do we have

$$\int_{SO(2)} \left| \widehat{\chi}_{P_N}(\rho\Theta) \right| \, d\theta \leq c \, \rho^{-2} \log \rho$$

for every $\rho \geq 2$.

6) Let C be a convex planar body. Prove that

$$\limsup_{\rho \to +\infty} \left(\rho^2 \log^{-1} \rho \int_{SO(2)} \left| \widehat{\chi}_C(\rho\Theta) \right| \, d\theta \right) > 0 \, .$$

7) Is the partition of unity necessary in the proof of Theorem 8.25?

9

Integer points and exponential sums

In this chapter we shall apply a theory developed by van der Corput [64, 65, 84, 100, 113, 123] to improve Dirichlet's estimate (see Theorem 2.14) for the divisor problem:

$$\sum_{1 \le n \le R} d(n) = R \log R + (2\gamma - 1) R + O\left(R^{1/2}\right) .$$

We shall obtain a result proved by Voronoï in 1903 [177] (see [84, 113, 123]).

9.1 Preliminary results

Voronoï's result is contained in the following theorem.

Theorem 9.1 (Voronoï)

$$\sum_{1 \le n \le R} d(n) = R \log R + (2\gamma - 1) R + O\left(R^{1/3} \log R\right) .$$

The proof will be given at the end of this chapter.

We first improve the estimate (2.11) for the difference

$$\sum_{1 \le n \le R} \frac{1}{n} - \log R .$$

Lemma 9.2

$$\sum_{1 \le n \le R} \frac{1}{n} = \log R + \gamma - \frac{s(R)}{R} + O\left(R^{-2}\right) ,$$

where $s(R)$ is the sawtooth function (4.33).

183

Proof   First we claim that for every $R \geq 1$ we have

$$\sum_{1 \leq n \leq R} \frac{1}{n} = \frac{1}{2} + \log R - \frac{s(R)}{R} - \int_1^R \frac{s(x)}{x^2} dx . \qquad (9.1)$$

Indeed, the terms in (9.1) are, as functions of R, continuous on $[1, \infty) \setminus \mathbb{Z}$ and continuous from the right in $\mathbb{Z}$, where they have jump discontinuities. Then it is enough to prove (9.1) for $R \notin \mathbb{Z}$. Since $s'(x) = 1$ for every $x \notin \mathbb{Z}$, we have

$$\sum_{1 \leq n \leq R} \frac{1}{n}$$

$$= 1 + \sum_{n=2}^{[R]} \int_{n-1}^{n} \frac{1}{n} dx$$

$$= 1 + \int_1^{[R]} \frac{1}{x} dx + \sum_{n=2}^{[R]} \int_{n-1}^{n} \left(\frac{1}{n} - \frac{1}{x} \right) dx$$

$$= 1 + \log R - \int_{[R]}^{R} \frac{1}{x} dx + \sum_{n=2}^{[R]} \int_{n-1}^{n} \left(\frac{1}{n} - \frac{1}{x} \right) s'(x) dx$$

$$= 1 + \log R - \int_{[R]}^{R} \frac{s'(x)}{x} dx + \sum_{n=2}^{[R]} \left\{ \left[s(x) \left(\frac{1}{n} - \frac{1}{x} \right) \right]_{x=(n-1)^+}^{x=n^-} - \int_{n-1}^{n} \frac{s(x)}{x^2} dx \right\}$$

$$= 1 + \log R - \left[\frac{s(x)}{x} \right]_{x=[R]^+}^{x=R} - \int_{[R]}^{R} \frac{s(x)}{x^2} dx + \frac{1}{2} \sum_{n=2}^{[R]} \left(\frac{1}{n} - \frac{1}{n-1} \right)$$

$$- \int_1^{[R]} \frac{s(x)}{x^2} dx$$

$$= \frac{1}{2} + \log R - \frac{s(R)}{R} - \int_1^R \frac{s(x)}{x^2} dx .$$

Hence (9.1) is proved. Since $\gamma = \lim_{R \to +\infty} \left(\sum_{n=1}^R \frac{1}{n} - \log R \right)$, (9.1) implies

$$\gamma = \frac{1}{2} - \int_1^{+\infty} \frac{s(x)}{x^2} dx .$$

Then (9.1) can be written as

$$\sum_{1 \leq n \leq R} \frac{1}{n} = \log R - \frac{s(R)}{R} + \gamma + \int_R^{+\infty} \frac{s(x)}{x^2} dx .$$

Let $S(x) := \int_0^x s(u) du$. Since $|S(x)| \leq 1/8$ for every $x \in \mathbb{R}$, we have

$$\int_R^{+\infty} \frac{s(x)}{x^2} dx = \left[\frac{S(x)}{x^2} \right]_{x=R}^{x=+\infty} + 2 \int_R^{+\infty} \frac{S(x)}{x^3} dx = O\left(R^{-2} \right) .$$

This proves the lemma. □

Now we show another way to write the error in the Dirichlet divisor problem.

Lemma 9.3 *For every $R \geq 1$ we have*

$$\sum_{1 \leq n \leq R} d(n) - R \log R - R(2\gamma - 1) = -2 \sum_{n \leq \sqrt{R}} s\left(\frac{R}{n}\right) + O(1) . \qquad (9.2)$$

Proof We split the sum $\sum_{1 \leq n \leq R} d(n)$ as in the proof of Theorem 2.14:

$$\sum_{1 \leq n \leq R} d(n) = \sum_{mr \leq R} 1 = \sum_{m \leq \sqrt{R}} \sum_{r \leq R/m} 1 + \sum_{r \leq \sqrt{R}} \sum_{m \leq R/r} 1 - \sum_{m \leq \sqrt{R}} \sum_{r \leq \sqrt{R}} 1$$

$$:= 2T - V .$$

By Lemma 9.2 we have

$$T = \sum_{m \leq \sqrt{R}} \left[\frac{R}{m}\right] = \sum_{m \leq \sqrt{R}} \left(\frac{R}{m} - s\left(\frac{R}{m}\right) - \frac{1}{2}\right)$$

$$= R\left(\frac{1}{2}\log R + \gamma\right) - \sqrt{R}\, s(\sqrt{R}) - \sum_{m \leq \sqrt{R}} s\left(\frac{R}{m}\right) - \frac{1}{2}\sqrt{R} + O(1) .$$

Since

$$V = \left[\sqrt{R}\right]^2 = \left(\sqrt{R} - s\left(\sqrt{R}\right) - \frac{1}{2}\right)^2 = R - 2\sqrt{R}\, s(\sqrt{R}) - \sqrt{R} + O(1) ,$$

we obtain (9.2). □

Lemma 9.4 *Let $f : \mathbb{N} \to \mathbb{R}$. Then there is $c > 0$, independent of f, such that for every interval $[a, b]$, having length at least 1, and for every $y > 0$ we have*

$$\left|\sum_{a \leq m \leq b} s(f(m))\right| \leq c\,\frac{b-a}{y} + c \sum_{n=1}^{+\infty} \min\left(\frac{y}{n^2}, \frac{1}{n}\right) \left|\sum_{a \leq m \leq b} e^{2\pi i n f(m)}\right| . \qquad (9.3)$$

Proof Let $x \in \mathbb{R}$ and $t > 0$. The graph of the sawtooth function $s(x)$ (see (4.33)) shows that

$$\frac{s(x) - s(x - t)}{t} \leq 1 ,$$

hence $s(x) \leq s(x - t) + t$. Then (4.34) implies

$$s(f(m)) \leq y \int_0^{y^{-1}} (s(f(m) - t) + t)\, dt = y \int_0^{y^{-1}} s(f(m) - t)\, dt + \frac{1}{2y}$$

$$= y \int_0^{y^{-1}} \left(-\frac{1}{\pi} \sum_{n=1}^{+\infty} \frac{1}{n} \sin(2\pi n (f(m) - t))\right) dt + \frac{1}{2y}$$

for every $y > 0$. Lemma 4.20 and the dominated convergence theorem imply

$$\sum_{a\leq m\leq b} s(f(m)) \leq \sum_{a\leq m\leq b}\left(-\frac{y}{\pi}\,\mathrm{Im}\left(\sum_{n=1}^{+\infty}\frac{1}{n}e^{2\pi i n f(m)}\int_0^{y^{-1}}e^{-2\pi i n t}\,dt\right) + \frac{1}{2y}\right)$$

$$\leq c\,\frac{b-a}{y} + c\sum_{n=1}^{+\infty}\left(\left|\frac{y}{n^2}\left(e^{-2\pi i n/y}-1\right)\right|\left|\sum_{a\leq m\leq b}e^{2\pi i n f(m)}\right|\right).$$

By (6.19) we have

$$\left|\frac{y}{n^2}\left(e^{-2\pi i n/y}-1\right)\right| \leq \frac{2y}{n^2}, \qquad \left|\frac{y}{n^2}\left(e^{-2\pi i n/y}-1\right)\right| \leq \frac{2\pi}{n}.$$

In a similar way, we start from the inequality $s(x+t)-t \leq s(x)$ and prove the estimate from below. Then (9.3) is proved. $\qquad\square$

9.2 Van der Corput's method and the Dirichlet divisor problem

Given a function $f : \mathbb{R} \to \mathbb{R}$ we call an *exponential sum* any sum

$$\sum_{n=a}^{b} e^{2\pi i f(n)}.$$

We have already seen a result (Theorem 6.18) where a regularity property of f gives a bound for an exponential sum. The following results are due to van der Corput.

Theorem 9.5 (van der Corput)   *Let $f \in C^2([a,b])$. Then there exists $c > 0$, independent of f such that if $|f''(x)| \geq \lambda_2 > 0$ on $[a,b]$, then*

$$\left|\sum_{a<m\leq b} s(f(m))\right| \leq c\,\frac{|f'(b)-f'(a)|}{\lambda_2^{2/3}} + c\,\frac{1}{\lambda_2^{1/2}}.$$

The proof depends on a few lemmas.

Lemma 9.6   *Let $f \in C^1([a,b])$ be a convex function such that $|f'(x)| \geq \lambda$ on $[a,b]$, then*

$$\left|\int_a^b e^{2\pi i f(x)}\,dx\right| \leq \lambda^{-1}.$$

Proof   Arguing as in (8.34) we obtain

$$\left|\int_a^b e^{2\pi i f(x)}dx\right| = \left|\int_a^b \frac{1}{2\pi i f'(x)}\frac{d}{dx}\left(e^{2\pi i f(x)}\right)dx\right|$$

$$= \left| \left[\frac{1}{2\pi i f'(x)} e^{2\pi i f(x)} \right]_a^b - \int_a^b \frac{d}{dx} \left(\frac{1}{2\pi i f'(x)} \right) e^{2\pi i f(x)} \, dx \right|$$

$$\leq \frac{1}{2\pi} \left(\left| \frac{1}{f'(b)} \right| + \left| \frac{1}{f'(a)} \right| \right) + \frac{1}{2\pi} \int_a^b \left| \frac{d}{dx} \left(\frac{1}{f'(x)} \right) \right| \, dx$$

$$\leq \frac{1}{\pi} \lambda^{-1} + \frac{1}{2\pi} \left| \int_a^b \frac{d}{dx} \left(\frac{1}{f'(x)} \right) dx \right| = \frac{1}{\pi} \lambda^{-1} + \frac{1}{2\pi} \left| \frac{1}{f'(b)} - \frac{1}{f'(a)} \right| < \lambda^{-1} ,$$

since $f'(x)$ is monotonic. $\qquad\square$

Lemma 9.7 *Let $f \in C^1 ([a, b])$ be a convex function such that $|f'(x)| \leq c < 1$ on $[a, b]$. Then*

$$\sum_{a \leq n \leq b} e^{2\pi i f(n)} = \int_a^b e^{2\pi i f(x)} \, dx + O(1) .$$

Proof We may assume that $b - a > 1$. Then Lemma 6.17 implies

$$\sum_{a \leq n \leq b} e^{2\pi i f(n)} = \int_a^b e^{2\pi i f(x)} \, dx + 2\pi i \int_a^b s(x) f'(x) e^{2\pi i f(x)} \, dx + O(1) .$$

By (4.34), Lemma 4.20 and the dominated convergence theorem we have

$$\int_a^b s(x) f'(x) e^{2\pi i f(x)} \, dx = -\frac{1}{\pi} \int_a^b \sum_{n=1}^{+\infty} \frac{1}{n} \sin(2\pi n x) f'(x) e^{2\pi i f(x)} \, dx$$

$$= \frac{i}{2\pi} \sum_{n \neq 0} \frac{1}{n} \int_a^b f'(x) e^{2\pi i (f(x) + nx)} \, dx .$$

Since f is convex and $|f'(x)| \leq c < 1$, then for every $n \neq 0$ the functions

$$\frac{f'(x)}{f'(x) + n}$$

exist and are monotonic. Therefore

$$\left| \int_a^b f'(x) e^{2\pi i (f(x) + nx)} \, dx \right| = \frac{1}{2\pi} \left| \int_a^b \frac{f'(x)}{f'(x) + n} \frac{d}{dx} \left(e^{2\pi i (f(x) + nx)} \right) dx \right|$$

$$\leq \frac{1}{2\pi} \left| \left[e^{2\pi i (f(x) + nx)} \frac{f'(x)}{f'(x) + n} \right]_a^b \right| + \frac{n}{2\pi} \left| \int_a^b e^{2\pi i (f(x) + nx)} \frac{f''(x)}{(f'(x) + n)^2} \, dx \right|$$

$$\leq \frac{c_1}{n} + \frac{c_2}{n} \int_a^b f''(x) \, dx \leq \frac{c_3}{n} .$$

Hence

$$\left| \int_a^b s(x) f'(x) e^{2\pi i f(x)} \, dx \right| \leq c_4 \sum_{n=1}^{+\infty} \frac{1}{n^2} = c_5 .$$

$\qquad\square$

Lemma 9.8 *Let $f \in C^2([a,b])$ satisfy $|f''(x)| \geq \lambda_2 > 0$ on $[a,b]$. Then*

$$\left| \int_a^b e^{2\pi i f(x)} dx \right| \leq \frac{4}{\sqrt{\lambda_2}} .$$

Proof We may assume that $f(x)$ is convex and let $[a,b] = I_1 \cup I_2$, where

$$I_1 := \left\{ x \in [a,b] : |f'(x)| \leq \lambda_2^{1/2} \right\}, \qquad I_2 := \left\{ x \in [a,b] : |f'(x)| > \lambda_2^{1/2} \right\} .$$

By convexity, I_1 is empty or it is an interval, while I_2 is the union of at most two intervals. Then Lemma 9.6 implies

$$\left| \int_{I_2} e^{2\pi i f(x)} dx \right| \leq \frac{2}{\sqrt{\lambda_2}} .$$

Now let $I_1 := [\alpha, \beta]$, then the mean value theorem implies

$$(\beta - \alpha)\lambda_2 \leq f'(\beta) - f'(\alpha) \leq 2\sqrt{\lambda_2} . \tag{9.4}$$

Hence

$$\left| \int_{I_1} e^{2\pi i f(x)} dx \right| \leq \beta - \alpha \leq \frac{2}{\sqrt{\lambda_2}} .$$

$\square$

Lemma 9.9 *Let $f \in C^2([a,b])$. Then there exists $c > 0$, independent of f, such that if $|f''(x)| \geq \lambda_2 > 0$ on $[a,b]$, then*

$$\left| \sum_{a<n\leq b} e^{2\pi i f(n)} \right| \leq c \, \frac{|f'(b) - f'(a)| + 1}{\sqrt{\lambda_2}} .$$

Proof We may assume that $f(x)$ is convex. The case $\lambda_2 \geq 1$ is easy, since arguing as in (9.4) we obtain

$$\left| \sum_{a<n\leq b} e^{2\pi i f(n)} \right| \leq b - a \leq (b-a)\sqrt{\lambda_2} \leq \frac{f'(b) - f'(a)}{\sqrt{\lambda_2}} .$$

As for the case $\lambda_2 < 1$, we consider an interval $(\alpha, \beta] \subseteq (a,b]$ and an integer h such that $|f'(x) - h| \leq 1/2$ for every $x \in (\alpha, \beta]$. Then Lemma 9.7 implies

$$\sum_{\alpha<n\leq\beta} e^{2\pi i f(n)} = \sum_{\alpha<n\leq\beta} e^{2\pi i (f(n)-hn)} = \int_\alpha^\beta e^{2\pi i (f(x)-hx)} dx + O(1) .$$

Lemma 9.8 gives

$$\left| \sum_{\alpha\leq n\leq\beta} e^{2\pi i f(n)} \right| \leq c_1 \frac{1}{\sqrt{\lambda_2}} .$$

The proof is complete because the interval $(a,b]$ can be subdivided into at most $O(f'(b) - f'(a) + 1)$ intervals of the previous type. $\square$

Proof of Theorem 9.5 We apply Lemma 9.9 to the function $m \mapsto nf(m)$. Then we obtain, for every positive integer n,

$$\left| \sum_{a<m\leq b} e^{2\pi i n f(m)} \right| \leq c \, \frac{n^{1/2} |f'(b) - f'(a)|}{\sqrt{\lambda_2}} + c \, \frac{1}{\sqrt{n\lambda_2}} .$$

Therefore (9.3) gives

$$\left| \sum_{a<m\leq b} s(f(m)) \right|$$

$$\leq c \, \frac{b-a}{y} + c \sum_{n=1}^{+\infty} \min\left(\frac{y}{n^2}, \frac{1}{n}\right) \left(\frac{n^{1/2} |f'(b) - f'(a)|}{\sqrt{\lambda_2}} + \frac{1}{\sqrt{n\lambda_2}} \right)$$

$$\leq c \, \frac{b-a}{y} + c \sum_{n=1}^{y} \frac{1}{n} \left(\frac{n^{1/2} |f'(b) - f'(a)|}{\sqrt{\lambda_2}} + \frac{1}{\sqrt{n\lambda_2}} \right)$$

$$+ c \sum_{n>y} \frac{y}{n^2} \left(\frac{n^{1/2} |f'(b) - f'(a)|}{\sqrt{\lambda_2}} + \frac{1}{\sqrt{n\lambda_2}} \right) .$$

We choose $y = \lambda_2^{-1/3}$. If $\lambda_2 \leq 1$ we have

$$c \left| \sum_{a<m\leq b} s(f(m)) \right| \leq (b-a) \lambda_2^{1/3} + \frac{|f'(b) - f'(a)|}{\sqrt{\lambda_2}} \sum_{n=1}^{\lambda_2^{-1/3}} \frac{1}{\sqrt{n}} + \frac{1}{\sqrt{\lambda_2}}$$

$$+ \frac{|f'(b) - f'(a)|}{\lambda_2^{5/6}} \sum_{n=\lambda_2^{-1/3}}^{+\infty} \frac{1}{n^{3/2}} + \frac{1}{\lambda_2^{5/6}} \sum_{n=\lambda_2^{-1/3}}^{+\infty} \frac{1}{n^{5/2}}$$

$$\leq \frac{|f'(b) - f'(a)|}{\lambda_2^{2/3}} + \frac{1}{\sqrt{\lambda_2}} ,$$

since the mean value theorem gives

$$\left| f'(b) - f'(a) \right| \geq (b-a) \lambda_2 .$$

If $\lambda_2 > 1$ we have

$$c \left| \sum_{a<m\leq b} s(f(m)) \right| \leq (b-a) \lambda_2^{1/3} + \frac{|f'(b) - f'(a)|}{\lambda_2^{5/6}} \sum_{n=1}^{+\infty} \frac{1}{n^{3/2}} + \frac{1}{\lambda_2^{5/6}} \sum_{n=1}^{+\infty} \frac{1}{n^{5/2}}$$

$$\leq \frac{|f'(b) - f'(a)|}{\lambda_2^{2/3}} + \frac{1}{\sqrt{\lambda_2}} .$$

$\square$

Corollary 9.10   *Let $f \in C^2((a,b])$, with $f''(x)$ monotonic and of constant sign. Then there exists $c > 0$, independent of f, such that*

$$\left| \sum_{a < m \le b} s(f(m)) \right| \le c \int_a^b |f''(x)|^{1/3} \, dx + c \frac{1}{|f''(a)|^{1/2}} + c \frac{1}{|f''(b)|^{1/2}} \, .$$

Proof   We first assume that $f''(x)$ is positive and increasing. Let

$$N = \left[\frac{\log |f''(b)| - \log |f''(a)|}{\log 2} \right] .$$

We subdivide $(a,b]$ into $N + 1$ intervals $\left(x_j, x_{j+1} \right]$ $(j = 0, 1, \dots, N)$ such that $x_0 = a$, $x_{N+1} = b$ and

$$2^j f''(a) \le f''(x) \le 2^{j+1} f''(a) , \qquad \text{if } x_j \le x \le x_{j+1} .$$

Then we apply Theorem 9.5 to each of the above intervals to obtain

$$\left| \sum_{a < m \le b} s(f(m)) \right| \le c \left| \sum_{j=0}^N \sum_{x_j \le m < x_{j+1}} s(f(m)) \right|$$

$$\le c_1 \sum_{j=0}^N \frac{f'(x_{j+1}) - f'(x_j)}{(2^j f''(a))^{2/3}} + c_1 \sum_{j=0}^N \frac{1}{(2^j f''(a))^{1/2}}$$

$$\le c_1 \sum_{j=0}^N \frac{1}{(2^j f''(a))^{2/3}} \int_{x_j}^{x_{j+1}} f''(x) \, dx + c_2 \frac{1}{f''(a)^{1/2}}$$

$$\le c_1 \sum_{j=0}^N \int_{x_j}^{x_{j+1}} f''(x)^{1/3} \, dx + c_2 \frac{1}{f''(a)^{1/2}} \, .$$

We treat the case when $|f''(x)|$ is decreasing in a similar way. $\square$

Proof of Theorem 9.1   By Lemma 9.3 it is enough to prove that

$$\left| \sum_{1 \le n \le \sqrt{R}} s\left(\frac{R}{n} \right) \right| \le c R^{1/3} \log R \, .$$

By applying Corollary 9.10 to the function $f(x) = R/x$ we obtain

$$\left| \sum_{1 \le n \le \sqrt{R}} s(f(n)) \right| \le c_1 R^{1/3} \int_1^{\sqrt{R}} \frac{1}{x} \, dx + c_1 R^{1/4} \le c R^{1/3} \log R \, .$$

$\square$

The Dirichlet divisor problem has a long history. As for the circle problem we define

$$\theta := \inf \left\{ \alpha \in \mathbb{R} : \sum_{1 \le n \le R} d(n) - R \log R - (2\gamma - 1) R = O(R^{\alpha}) \right\} .$$

The following estimates have been proved so far:

$\theta \le 1/2$ Dirichlet (1849)

$\theta \le 1/3$ Voronoï (1903)

$\theta \le 0.33$ van der Corput (1922)

$\theta \le 27/82 = 0.32927 \cdots$ van der Corput (1928)

$\theta \le 15/46 = 0.32609 \cdots$ Chih Tsung-tao (1950)

$\theta \le 12/37 = 0.32432 \cdots$ Kolesnik (1969)

$\theta \le 346/1067 = 0.32427 \cdots$ Kolesnik (1973)

$\theta \le 35/108 = 0.32407 \cdots$ Kolesnik (1982)

$\theta \le 139/429 = 0.32401 \cdots$ Kolesnik (1985)

$\theta \le 7/22 = 0.31818 \cdots$ Iwaniec and Mozzochi (1987)

$\theta \le 23/73 = 0.31506 \cdots$ Huxley (1993)

$\theta \le 131/416 = 0.31490 \cdots$ Huxley (2003).

The conjecture $\theta = 1/4$ is widely believed to be true. Hardy [89] and Landau [116] (see also [174]) proved that for no positive constant c do we have the upper bound

$$\left| \sum_{1 \le n \le R} d(n) - R \log R - (2\gamma - 1) R \right| \le c R^{1/4}$$

for every $R \ge 1$.

Comparing the above list with the list in Section 8.3, we see that the Dirichlet divisor problem and the Gauss circle problem seem to have a sort of parallel development. We recall that we have proved Sierpinski's result (Theorem 8.17) using a 2-dimensional Fourier analytic argument. It is therefore natural to ask whether Fourier analysis on $\mathbb{T}^2$ can give a non-trivial result for the divisor problem also. The answer is affirmative, but there is a key difference between the above two problems, since the area under the hyperbola is infinite. Positive results have been obtained recently by applying the distribution theory and Fourier transforms [57, 163].

Exercises

1) Let k be an integer ≥ 2. Let $f \in C^k((a,b))$ satisfy $\left|f^{(k)}(x)\right| \geq 1$ for every $x \in (a,b)$. Prove that for every $\rho \geq 1$ we have

$$\left|\int_a^b e^{2\pi i \rho f(x)}\, dx\right| \leq c_k\, \rho^{-1/k}\;.$$

2) Let $f(x)$ be smooth on (a,b) with $f'(x) \neq 0$ for every $x \in [a,b]$. Let $\varphi(x)$ be smooth on (a,b) with compact support in (a,b). Prove that for every $\rho \geq 1$ and $N \geq 1$ we have

$$\left|\int_a^b e^{2\pi i \rho f(x)} \varphi(x)\, dx\right| \leq c_N\, \rho^{-N}\;.$$

3) Let $f, g \in C^2([a,b])$. Let $\frac{g(x)}{f'(x)}$ be monotonic and assume that $\left|\frac{g(x)}{f'(x)}\right| \leq \lambda$ on $[a,b]$. Prove that

$$\int_a^b e^{2\pi i f(x)} g(x)\, dx \leq c\, \frac{1}{\lambda}\;.$$

4) Let $f \in C^2([a,b])$, with $f''(x)$ positive and monotonic on $[a,b]$. Let $1 \leq r_{\max} < \infty$ be the maximum of the radius of curvature of the graph of the curve $y = f(x)$. Prove that

$$\left|\sum_{a < m \leq b} s(f(m))\right| \leq c\, (r_{\max})^{2/3}\;.$$

10

Geometric discrepancy and decay of Fourier transforms

In this chapter we introduce a fundamental result of Roth which has given rise to the so-called 'geometric discrepancy', a field of research where finite sequences of points are tested against suitable families of sets. Then we shall introduce a theorem, proved independently by Beck and Montgomery, which gives a lower L^2 bound for a discrepancy associated with convex bodies. The known proofs of this result depend on suitable estimates of the average decay of Fourier transforms of characteristic functions of convex bodies. We have already encountered this topic in Section 8.4 and we shall investigate it again in Section 10.3. We shall end this chapter by showing that the result of Beck and Montgomery is sharp.

10.1 Choosing N points in a cube

Let N be a large positive integer, and let $A_N = \{t(j)\}_{j=1}^N$ be a set consisting of N points in the d-dimensional cube $[0, 1)^d$, treated as a torus. We consider the problem of choosing the points $t(1), t(2), \ldots, t(N)$ in such a way that the *discrepancy*

$$D_N = D(A_N) := \sup_{I \subseteq \mathbb{T}^d} |\mathrm{card}\, (A_N \cap I) - N\, |I||$$

is as small as possible. Here the sup is on the d-dimensional intervals

$$I = [a_1, b_1] \times \ldots \times [a_d, b_d]$$

contained in the torus $\mathbb{T}^d$.

When $d = 1$ it is natural to choose the N points at distance $1/N$:

$$\{t(j)\}_{j=1}^N = \left\{\frac{1}{N}, \frac{2}{N}, \ldots, \frac{N}{N}\right\}. \tag{10.1}$$

We observe that the finite set $\{t(j)\}_{j=1}^{N}$ in (10.1) has discrepancy 1. Indeed, for every interval $[a, b]$, which we may assume to be contained in $[0, 1)$, let k be the integer satisfying

$$\frac{k}{N} \le b - a < \frac{k+1}{N} .$$

Then $[a, b]$ contains either k or $k + 1$ elements of A_N. Therefore $D_N = 1$.

For every sequence A_N of N elements in $\mathbb{T}^d$ we have $D(A_N) \ge 1$. Indeed, if we choose $t \in A_N$ and $\varepsilon > 0$ such that the interval $I_\varepsilon = [t - \varepsilon/2, t + \varepsilon/2]^d$ does not contain other points of A_N we have

$$\text{card}(A_N \cap I_\varepsilon) - N|I_\varepsilon| \ge 1 - N\varepsilon^d .$$

Since ε can be arbitrarily small we obtain $D(A_N) \ge 1$.

The idea of choosing the N points at a given distance cannot be applied to an infinite sequence in $\mathbb{T}$. Indeed, suppose that we have carefully chosen the first N points $t(1), t(2), \ldots, t(N)$. Then it is not clear how to select the next few points $t(N + 1), t(N + 2), \ldots$ Every choice of these latter points seems to violate the uniformity achieved through the first N points. The *van der Corput sequence*[1]

$$\{t(j)\}_{j=1}^{\infty} := \left\{ \frac{1}{2}, \frac{1}{4}, \frac{3}{4}, \frac{1}{8}, \frac{5}{8}, \frac{3}{8}, \frac{7}{8}, \frac{1}{16}, \frac{9}{16}, \frac{5}{16}, \frac{13}{16}, \frac{3}{16}, \frac{11}{16}, \frac{7}{16}, \frac{15}{16}, \ldots \right\}$$
(10.2)

is an attempt at choosing a 'good' infinite sequence inside the unit interval. The problem is that the partial sequences $\{t(j)\}_{j=1}^{N}$ in (10.2) are 'well distributed' only when we move from a denominator to the next one, but not for the intermediate values.[2] This may suggest, as conjectured by van der Corput [66], that for every infinite sequence $\{t(j)\}_{j=1}^{\infty} \subset \mathbb{T}$ we have $\sup_N D\left(\{t(j)\}_{j=1}^{N}\right) = +\infty$. This conjecture was proved in 1945 by van Aardenne-Ehrenfest [1, 2]. Her result was improved in 1954 by Roth [149] as a consequence of a theorem on the discrepancy of finite sequences in the unit square. In this way Roth started a new field, later called *geometric discrepancy* (see [14, 19, 43, 45, 48, 52, 73, 121]). Consider, for example, the van der Corput sequence (10.2). In order to see it as a sequence (and not only as the image of a sequence), we look at its graph:

---

[1] This simple and well-known sequence is defined as follows: for every $j = \sum a_k 2^k$ (written in base 2), let

$$t(j) := \sum a_k 2^{-k-1} .$$

[2] In fact we shall prove in Theorem 10.3 that the discrepancy of every infinite sequence is unbounded.

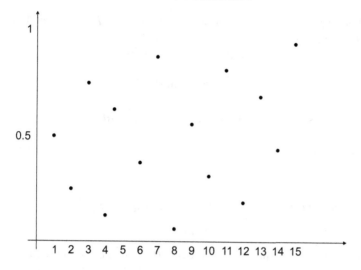

so we are led to study a distribution of points in two variables.

Let us point out a few differences between discrepancies in one and two variables. The intervals are of course the most familiar subsets of $[0, 1)$, but in the unit square we have several reasonable choices: discs, rectangles, polygons, convex bodies, ... The boundary of an interval can contain at most two points of the sequence, while the boundary of a planar set can contain many points and thereby have a great influence on the discrepancy (see the remarks at the beginning of Section 8.4). Also note that the analogue of the finite set in (10.1) should be $\left(\frac{1}{\sqrt{N}}\mathbb{Z}^2\right) \cap [0, 1)^2$, but this choice would require $\sqrt{N}$ to be an integer number.

10.2 Roth's theorem

Roth's result tests a finite sequence in $\mathbb{T}^2$ against a family of rectangles.

Theorem 10.1 (Roth) *There exists a constant $c > 0$ such that for every set A_N of N points in $\mathbb{T}^2$ we have*

$$\int_{\mathbb{T}^2} |\text{card}\,(A_N \cap I_x) - Nx_1x_2|^2 \; dx_1dx_2 \geq c \log(N) \, ,$$

where $I_x = [0, x_1] \times [0, x_2]$ for every $x = (x_1, x_2) \in [0, 1)^2$.

Then we deduce the following result.

Theorem 10.2   *There exists a constant $c > 0$ such that for every set A_N of N points in $\mathbb{T}^2$ we have*

$$D(A_N) \geq c \log^{1/2} (N) .$$

Proof of Theorem 10.2   By Roth's theorem we have

$$c \log^{1/2} (N) \leq \left\{ \int_{[0,1)^2} |\text{card} (A_N \cap I_x) - N x_1 x_2|^2 \, dx_1 dx_2 \right\}^{1/2}$$

$$\leq \sup_{x \in [0,1)^2} |\text{card} (A_N \cap I_x) - N |I_x|| \leq \sup_{I \subseteq \mathbb{T}^2} |\text{card} (A_N \cap I) - N |I|| = D(A_N) .$$

$\square$

Proof of Theorem 10.1   Let n be the positive integer satisfying

$$2^{n-1} < 2N \leq 2^n .$$

For every non-negative integer $j \leq n$, write $[0, 1)^2$ as the disjoint union of the 2^n rectangles

$$[0, 1)^2 = \bigcup_{\substack{0 \leq m_1 < 2^j \\ 0 \leq m_2 < 2^{n-j}}} B^j_{m_1, m_2} ,$$

where

$$B^j_{m_1, m_2} := \left[m_1 2^{-j}, (m_1 + 1) 2^{-j} \right) \times \left[m_2 2^{j-n}, (m_2 + 1) 2^{j-n} \right) .$$

The following figure depicts two cases:

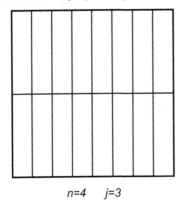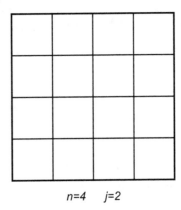

n=4   j=3   n=4   j=2

Now we subdivide every rectangle $B^j_{m_1, m_2}$ into four smaller rectangles

$$B' (j), B'' (j), B''' (j), B'''' (j)$$

and define the function $R_j : [0, 1]^2 \rightarrow \{\pm 1\}$ which in each rectangle $B^j_{m_1, m_2}$ takes values as in the following figure:

-1	+1
B'''(j)	B''''(j)
+1	-1
B'(j)	B''(j)

We recall that every $x \in [0, 1)^2$ belongs to a rectangle $B^j_{m_1,m_2}$. For every $0 \le j \le n$ we define the functions

$$f_j(x) := \begin{cases} R_j(x) & \text{if } x \in B^j_{m_1,m_2} \text{ and } B^j_{m_1,m_2} \cap A_N = \varnothing, \\ 0 & \text{otherwise,} \end{cases}$$

$$F(x) := \sum_{j=0}^{n} f_j(x).$$

The functions f_j are pairwise orthogonal. Indeed, if we assume $0 \le j < k \le n$, we can write $[0, 1)^2$ as the disjoint union of 2^{n+k-j} rectangles by subdividing the horizontal side into 2^k equal intervals and the vertical side into 2^{n-j} equal intervals. Let S be one of these 2^{n+k-j} rectangles. Then S is contained in a rectangle $B^j_{m_1,m_2}$. If $x \in S$ we have either $f_j(x)f_k(x) = 0$ or $f_j(x)f_k(x) = \pm 1$ in such a way that $\int_S f_j f_k = 0$. Then

$$\int_{\mathbb{T}^2} |F(x)|^2 \, dx = \int_{\mathbb{T}^2} \sum_{j=0}^{n} f_j(x) \sum_{k=0}^{n} f_k(x) \, dx = \sum_{j=0}^{n} \int_{\mathbb{T}^2} |f_j(x)|^2 \, dx \qquad (10.3)$$

$$\le n + 1 \le c \log N .$$

Now let

$$D(x) := \text{card} (A_N \cap I_x) - N x_1 x_2 . \qquad (10.4)$$

We are going to show that, for every j,

$$\left| \int_{\mathbb{T}^2} f_j(x) D(x) \, dx \right| \ge c_1 . \qquad (10.5)$$

From now on we shall write B_j in place of $B^j_{m_1,m_2}$. In order to prove (10.5) we observe that if B_j contains points of A_N, then $f_j|_{B_j} = 0$ and therefore $\int_{B_j} f_j(x) D(x) \, dx = 0$. If $B_j \cap A_N = \varnothing$, let $B'(j), B''(j), B'''(j), B''''(j)$ be as above. For each $x \in B'(j)$ we consider the rectangle I_{xyzw} (with side lengths 2^{-j-1} and 2^{j-n-1}) in the following figure:

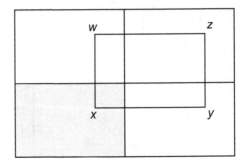

We write the vertices of I_{xyzw} as $x = (x_1, x_2)$, $y = (z_1, x_2)$, $z = (z_1, z_2)$, $w = (x_1, z_2)$. We recall that $B_j \cap A_N = \emptyset$. Then

$$\int_{B_j} f_j(u)D(u)du \tag{10.6}$$

$$= \int_{B'(j)} D(u)\,du - \int_{B''(j)} D(u)\,du + \int_{B'''(j)} D(u)\,du - \int_{B''''(j)} D(u)\,du$$

$$= \int_{B'(j)} D(x)\,dx - \int_{B'(j)} D(y)\,dx + \int_{B'(j)} D(z)\,dx - \int_{B'(j)} D(w)\,dx$$

$$= \int_{B'(j)} (\operatorname{card}(A_N \cap I_x) - Nx_1 x_2)\,dx - \int_{B'(j)} \left(\operatorname{card}\left(A_N \cap I_y\right) - Nz_1 x_2\right) dx$$

$$+ \int_{B'(j)} (\operatorname{card}(A_N \cap I_z) - Nz_1 z_2)\,dx - \int_{B'(j)} (\operatorname{card}(A_N \cap I_w) - Nx_1 z_2)\,dx$$

$$= \int_{B'(j)} \left(\operatorname{card}\left(A_N \cap I_{xyzw}\right) - N(z_2 - x_2)(z_1 - x_1)\right) dx = -N2^{-2n-4}\ ,$$

since $B_j \cap A_N = \emptyset$ and $\left|I_{xyzw}\right| = |B'(j)| = 2^{-n-2}$. Now observe that there are at least 2^{n-1} rectangles B_j which do not contain points of A_N, so that for these rectangles (10.6) holds true. Then, if we recall that f_j vanishes on the remaining rectangles, we have

$$\int_{\mathbb{T}^2} f_j(x)D(x)dx \le -N2^{-2n-4}2^{n-1} = -N2^{-n-5} \le -c_1\ .$$

Hence

$$\int_{\mathbb{T}^2} F(x)D(x)\,dx = \sum_{j=0}^{n} \int_{\mathbb{T}^2} f_j(x)D(x)\,dx \le -c_1\,(n+1) \le -c_2 \log(N)\ .$$

Then by (10.3) and the Cauchy–Schwarz inequality we obtain

$$\log(N) \le c_3 \left| \int_{\mathbb{T}^2} F(x)D(x)\,dx \right| \le c_3 \left\{ \int_{\mathbb{T}^2} |F(x)|^2\,dx \right\}^{1/2} \left\{ \int_{\mathbb{T}^2} |D(x)|^2\,dx \right\}^{1/2}$$

$$\leq c_4 \log^{1/2}(N) \left\{ \int_{\mathbb{T}^2} |D(x)|^2 \, dx \right\}^{1/2}$$

to complete the proof. □

Roth's theorem yields a proof of the van der Corput conjecture.

Theorem 10.3 *Let $t(j)$ be an infinite sequence with values in $\mathbb{T}$. Then for infinitely many values of N we have*

$$D_N = \sup_{I \subseteq \mathbb{T}} \left| \text{card}\left(\{t(j)\}_{j=1}^N \cap I\right) - N|I| \right| \geq c \log^{1/2}(N) \, ,$$

where the sup *is over all intervals $I \subseteq \mathbb{T}$.*

Proof Fix N and consider the finite sequence

$$\{p(j)\}_{j=1}^N = \left\{ \left(\frac{j}{N}, t(j) \right) \right\}_{j=1}^N \subset \mathbb{T}^2 \, .$$

By Roth's theorem we know that there exists $x = (x_1, x_2) \in [0, 1)^2$ such that the rectangle $I_x := [0, x_1] \times [0, x_2]$ satisfies

$$\left| \text{card}\left(\left\{ \left(\frac{j}{N}, t(j) \right) \right\}_{j=1}^N \cap I_x \right) - N x_1 x_2 \right| \geq c \log^{1/2}(N) \, . \qquad (10.7)$$

Observe that

$$\left| \text{card}\left(\{t(j)\}_{1 \leq j \leq N x_1} \cap [0, x_2] \right) - [N x_1] \, x_2 \right|$$
$$\geq \left| \text{card}\left(\left\{ \left(\frac{j}{N}, t(j) \right) \right\}_{j=1}^N \cap I_x \right) - N x_1 x_2 \right| - 1 \, .$$

Since x depends on N we cannot say that the integer sequence $[N x_1]$ fills the whole of $\mathbb{N}$ (and in fact this does not happen for the van der Corput sequence). Anyway, the sequence $N x_1$ diverges. Otherwise we would have $x_1 \leq c/N$ for infinitely many values of N. Then only a bounded number of points $p(j)$ would belong to I_x and $N x_1 x_2$ would be bounded too. This would contradict (10.7). □

The study of low discrepancy sequences in several variables (usually called *quasi-Monte Carlo methods*) is a basic instrument for the approximate computation of integrals in many variables. These integrals may appear in statistics, physics and mathematical finance (see e.g. [73, 104, 124, 131]).

10.3 Average decay of Fourier transforms

In this section we continue the study of the spherical averages of Fourier transforms of characteristic functions (see Section 8.4). The following results will be used both in lower and upper estimates for the L^2 discrepancy [30, 123]. They are also useful in the study of generalized Radon transforms [29, 31, 147] and other problems (see e.g. [36, 111]).

Lemma 10.4 *Let $C \subset \mathbb{R}^2$ be a convex body with characteristic function χ_C. Then there exist four positive constants $\alpha, \beta, \gamma, \delta$ such that for every $\rho \geq 1$ we have*

$$\alpha \rho^{-1} \leq \int_{\{\gamma\rho \leq |\xi| \leq \delta\rho\}} \left|\widehat{\chi_C}(\xi)\right|^2 d\xi \leq \beta\rho^{-1} . \tag{10.8}$$

Proof The upper bound is a direct consequence of Theorem 8.25. Here we give a direct proof. We show that

$$\int_{\{|\xi| \geq \rho\}} \left|\widehat{\chi_C}(\xi)\right|^2 d\xi \leq c\rho^{-1} . \tag{10.9}$$

In order to prove (10.9) it is enough to show that for every integer $k \geq 0$ we have

$$\int_{\{2^k \leq |\xi| \leq 2^{k+1}\}} \left|\widehat{\chi_C}(\xi)\right|^2 d\xi \leq c_1 2^{-k} . \tag{10.10}$$

By the convexity assumption there exist two positive constants c_2 and c_3 such that, for every small $h \in \mathbb{R}^2$,

$$c_2 |h| \leq |C \triangle (C + h)| \leq c_3 |h| , \tag{10.11}$$

where $A \triangle B = (A \backslash B) \cup (B \backslash A)$. Then, by Plancherel's theorem (8.4),

$$|h| \approx \int_{\mathbb{R}^2} |\chi_C(x+h) - \chi_C(x)|^2 dx = \int_{\mathbb{R}^2} \left|e^{2\pi i \xi \cdot h} - 1\right|^2 \left|\widehat{\chi_C}(\xi)\right|^2 d\xi . \tag{10.12}$$

Now we subdivide each annulus

$$\{2^k \leq |\xi| \leq 2^{k+1}\} := \bigcup_{j=1}^{V} A_j$$

into 'slices' A_j as follows. The number V does not depend on k and for each $j = 1, \ldots, V$ there exists a small $h_j \in \mathbb{R}^2$ such that $c_4 2^{-k} \leq |h_j| \leq c_5 2^{-k}$ and $\left|e^{2\pi i \xi \cdot h_j} - 1\right| \geq c_6$ for every $\xi \in A_j$. Then the upper estimate in (10.8) follows from (10.9). We now consider the lower estimate. Observe that (10.10) implies

$$\int_{\{|\xi| \leq \gamma\rho\}} |\xi|^2 \left|\widehat{\chi_C}(\xi)\right|^2 d\xi \leq c_7 + c_8 \sum_{k=1}^{[\log_2(\gamma\rho)]} \int_{2^k \leq |\xi| \leq 2^{k+1}} |\xi|^2 \left|\widehat{\chi_C}(\xi)\right|^2 d\xi \tag{10.13}$$

$$\le c_7 + c_9 \sum_{k=1}^{[\log_2(\gamma\rho)]} 2^{2k} \int_{2^k \le |\xi| \le 2^{k+1}} \left|\widehat{\chi_C}(\xi)\right|^2 d\xi$$

$$\le c_7 + c_{10} \sum_{k=1}^{[\log_2(\gamma\rho)]} 2^{2k} 2^{-k} \le c_{11}\gamma\rho \ .$$

Then (10.12), (6.19) and (10.13) imply

$$c_{12}|h| \le 4\pi^2 |h|^2 \int_{\{|\xi| \le \gamma\rho\}} |\xi|^2 \left|\widehat{\chi_C}(\xi)\right|^2 d\xi + 4 \int_{\{\gamma\rho \le |\xi| \le \delta\rho\}} \left|\widehat{\chi_C}(\xi)\right|^2 d\xi$$

$$+ 4 \int_{\{|\xi| \ge \delta\rho\}} \left|\widehat{\chi_C}(\xi)\right|^2 d\xi$$

$$\le c_{13} \left[\gamma\rho |h|^2 + \delta^{-1}\rho^{-1}\right] + 4 \int_{\{\gamma\rho \le |\xi| \le \delta\rho\}} \left|\widehat{\chi_C}(\xi)\right|^2 d\xi \ .$$

Finally, if we let $|h| = \rho^{-1}$ and the constants γ and δ are, respectively, suitably small and suitably large, we obtain

$$\int_{\{\gamma\rho \le |\xi| \le \delta\rho\}} \left|\widehat{\chi_C}(\xi)\right|^2 d\xi \ge \frac{c_{12}}{4}\rho^{-1} - \frac{c_{13}}{4}\left[\gamma\rho^{-1} + \delta^{-1}\rho^{-1}\right] \ge c_{14}\rho^{-1} \ .$$

$$\square$$

The above lemma shows that the estimate in Theorem 8.25 is best possible. Indeed, (10.8) readily implies

$$\limsup_{\rho \to +\infty} \rho^{3/2} \left\|\widehat{\chi_C}(\rho\cdot)\right\|_{L^2([0,2\pi))} > 0 \ .$$

Remark 10.5 The lower bound

$$\left\|\widehat{\chi_C}(\rho\cdot)\right\|_{L^2([0,2\pi))} \ge c\rho^{-3/2}$$

may fail. Indeed, let $Q = [-1/2, 1/2]^2$ be a unit square and let $k \in \mathbb{N}$. Then, for every $p > 1$, (8.11) implies

$$\int_0^{2\pi} \left|\widehat{\chi_Q}(k\cos\theta, k\sin\theta)\right|^p d\theta = 8 \int_0^{\pi/4} \left|\frac{\sin(\pi k \cos\theta)}{\pi k \cos\theta} \frac{\sin(\pi k \sin\theta)}{\pi k \sin\theta}\right|^p d\theta$$

$$\le c_p \frac{1}{k^{2p}} \int_0^{\pi/4} \left|\frac{\sin(\pi k \cos\theta)}{\sin\theta}\right|^p d\theta \le c_p' \frac{1}{k^{2p}} \int_0^{\pi/4} \left|\sin(2\pi k \sin^2(\theta/2))\right|^p \theta^{-p} d\theta$$

$$\le c_p'' \frac{1}{k^{2p}} \int_0^{k^{-1/2}} k^p \theta^p d\theta + c_p'' \frac{1}{k^{2p}} \int_{k^{-1/2}}^{\pi/4} \theta^{-p} d\theta \le c_p''' k^{-(3p+1)/2} \ ,$$

which is $o\left(k^{-p-1}\right)$ as $k \to +\infty$.

For a triangle we have the expected lower bound.

Lemma 10.6   *Let T be a triangle in $\mathbb{R}^2$ and let $1 < p \le +\infty$. Then there exists a constant $c_p > 0$ such that, for large ρ,*

$$\left\| \widehat{\chi_T}\,(\rho \cdot) \right\|_{L^p((0,2\pi))} \ge c\,\rho^{-1-1/p} \ .$$

Proof   We may assume that $1 < p < +\infty$. Let $\Theta = (\cos \theta, \sin \theta)$ and let

$$\omega(t) = \frac{-1}{2\pi i \rho} e^{-2\pi i \rho \Theta \cdot t}\, \Theta \ .$$

As in the proof of Theorem 8.25, we have $\operatorname{div}(\omega(t)) = e^{-2\pi i \rho \Theta \cdot t}$. Then the divergence theorem implies

$$\widehat{\chi_T}\,(\rho \Theta) = \int_T e^{-2\pi i \rho \Theta \cdot t}\, dt = \int_{\partial T} \omega(s) \cdot v(s)\, ds \ ,$$

where ds is the 1-dimensional measure on the three sides $\lambda_1, \lambda_2, \lambda_3$, while $v(s)$ is the outward unit vector at the point s. Since T is a triangle, $v(s)$ takes only three values: v_1, v_2, v_3.

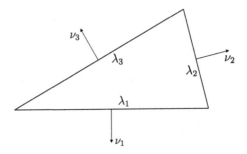

Then

$$\widehat{\chi_T}\,(\rho \Theta) = \frac{-\Theta \cdot v_1}{2\pi i \rho} \int_{\lambda_1} e^{-2\pi i \rho \Theta \cdot s}\, ds - \frac{\Theta \cdot v_2}{2\pi i \rho} \int_{\lambda_2} e^{-2\pi i \rho \Theta \cdot s}\, ds$$

$$- \frac{\Theta \cdot v_3}{2\pi i \rho} \int_{\lambda_3} e^{-2\pi i \rho \Theta \cdot s}\, ds$$

$$:= A(\rho, \Theta) + B(\rho, \Theta) + C(\rho, \Theta) \ .$$

We may assume that

$$\lambda_1 = \left\{ (s, 0) : -\frac{1}{2} \le s \le \frac{1}{2} \right\}$$

is the base of T. It is enough to prove that, for a given δ, we have

$$\int_{\frac{\pi}{2}-\delta}^{\frac{\pi}{2}+\delta} \left| \widehat{\chi_T}\,(\rho \Theta) \right|^p\, d\theta \ge c\,\rho^{-p-1} \ .$$

We show that

$$\int_{\frac{\pi}{2}-\delta}^{\frac{\pi}{2}+\delta} |A(\rho, \Theta)|^p \, d\theta \geq c \, \rho^{-p-1} .$$

Indeed, since $|\Theta \cdot \nu_1| = |\sin \theta|$ we have, for large ρ and suitable constants,

$$\int_{\frac{\pi}{2}-\delta}^{\frac{\pi}{2}+\delta} |A(\rho, \Theta)|^p \, d\theta = c_p \frac{1}{\rho^p} \int_{\frac{\pi}{2}-\delta}^{\frac{\pi}{2}+\delta} \left| \sin \theta \int_{-1/2}^{1/2} e^{-2\pi i \rho s \cos \theta} - ds \right|^p d\theta$$

$$= c_p \frac{1}{\rho^p} \int_{\frac{\pi}{2}-\delta}^{\frac{\pi}{2}+\delta} \left| \frac{\sin(\pi \rho \cos \theta)}{\pi \rho \cos \theta} \sin \theta \right|^p d\theta$$

$$\geq c'_p \frac{1}{\rho^{p+1}} \int_{\frac{\pi}{2}}^{\frac{\pi}{2}+c_1/\rho} \left(\frac{\sin(\pi \rho \cos \theta)}{\pi \rho \cos \theta} \sin \theta \right)^p d\theta$$

$$\geq c'_p \frac{1}{\rho^{p+1}} \int_0^{\pi/4} \frac{(\sin(x))^p}{x^p} \, dx \geq c''_p \, \rho^{-p-1} .$$

As for $B(\rho, \Theta)$ and $C(\rho, \Theta)$, the restriction $\left| \theta - \frac{\pi}{2} \right| \leq \delta$ and a similar computation lead us to evaluate two integrals of the form

$$\frac{1}{\rho^p} \int_{c_3}^{c_4} \left| \frac{\sin(2\pi \rho x)}{\rho x} \right|^p dx ,$$

where $0 < c_3 < c_4$. Then

$$\int_{\frac{\pi}{2}-\delta}^{\frac{\pi}{2}+\delta} |B(\rho, \Theta)|^p \, d\theta + \int_{-\frac{\pi}{2}-\delta}^{-\frac{\pi}{2}+\delta} |C(\rho, \Theta)|^p \, d\theta \leq c_5 \rho^{-2p} .$$

$\square$

10.4 Irregularities of distribution for convex bodies

In this section we use the results on the average decay of Fourier transforms to study the discrepancy of a finite sequence $\{t(j)\}_{j=1}^N \subset \mathbb{T}^2$ with respect to a family $\{B_\alpha\}$ of subsets of $\mathbb{T}^2$. If the family is too large or too small, then the problem may have little interest. It is meaningful to consider families of discs, rectangles, polygons, convex bodies, ... and look for finite sequences which yield a small discrepancy for the whole family.

Let C be a convex body with diameter less than 1 and let $\varepsilon \sigma^{-1}(C) - t$ be the rotated, dilated and translated copy of C, where $\sigma \in SO(2)$, $0 < \varepsilon \leq 1$ and $t \in \mathbb{T}^2$. Note that the assumption on the diameter of C makes the projection of $\varepsilon \sigma(C) - t$ from $\mathbb{R}^2$ on $\mathbb{T}^2$ injective. We define the discrepancies

$$D_N^{C,\{t(1),\dots,t(N)\}} := -N\,|C| + \sum_{j=1}^{N} \chi_C(t(j)),\tag{10.14}$$

$$D_N^{C,\{t(1),\dots,t(N)\}}(t) = D_N(t) := -N\,|C| + \sum_{j=1}^{N} \chi_{C-t}(t(j)),$$

$$D_N^{C,\{t(1),\dots,t(N)\}}(\sigma,t) = D_N(\sigma,t) := -N\,|C| + \sum_{j=1}^{N} \chi_{\sigma^{-1}(C)-t}(t(j)),$$

$$D_N^{C,\{t(1),\dots,t(N)\}}(\varepsilon,\sigma,t) = D_N(\varepsilon,\sigma,t) := -N\varepsilon^2\,|C| + \sum_{j=1}^{N} \chi_{\varepsilon\sigma^{-1}(C)-t}(t(j)).$$

When no confusion will arise we simply write $D_N(t)$ in place of $D_N(\sigma,t)$ or $D_N(\varepsilon,\sigma,t)$.

We show that the function $t \mapsto D_N^C(\varepsilon,\sigma,t) = D_N(t)$ has Fourier series

$$\sum_{k\neq 0} \left(\sum_{j=1}^{N} e^{2\pi i k \cdot t(j)} \right) \varepsilon^2\, \widehat{\chi_C}(\varepsilon\sigma(k))\, e^{2\pi i k \cdot t},\tag{10.15}$$

where we note that here $\widehat{\chi_C}$ is a Fourier transform on $\mathbb{R}^2$. We compute $\widehat{D_N}(k)$. Since

$$\int_{\mathbb{T}^2} \left(\sum_{j=1}^{N} \chi_{\varepsilon\sigma^{-1}(C)-t}(t(j)) \right) dt = \sum_{j=1}^{N} \int_{\mathbb{T}^2} \chi_{\varepsilon\sigma^{-1}(C)}(t(j)+t)\, dt$$

$$= \sum_{j=1}^{N} \int_{\mathbb{T}^2} \chi_{\varepsilon\sigma^{-1}(C)}(t)dt = N\varepsilon^2\,|C|$$

we have $\widehat{D_N}(0) = 0$. For $k \neq 0$ we have

$$\widehat{D_N}(k) = \int_{\mathbb{T}^2} \left(\sum_{j=1}^{N} \chi_{\varepsilon\sigma^{-1}(C)-t}(t(j)) - N\varepsilon^2\,|C| \right) e^{-2\pi i k \cdot t}\, dt$$

$$= \sum_{j=1}^{N} \int_{\mathbb{T}^2} \chi_{\varepsilon\sigma^{-1}(C)}(t(j)+t) e^{-2\pi i k \cdot t}\, dt = \sum_{j=1}^{N} e^{2\pi i k \cdot t(j)} \int_{\mathbb{T}^2} \chi_{\varepsilon\sigma^{-1}(C)}(u) e^{-2\pi i k \cdot u}\, du$$

$$= \sum_{j=1}^{N} e^{2\pi i k \cdot t(j)}\, \widehat{\chi_{\varepsilon\sigma^{-1}(C)}}(k) = \sum_{j=1}^{N} e^{2\pi i k \cdot t(j)} \varepsilon^2\, \widehat{\chi_C}(\varepsilon\sigma(k)).$$

We now recall the Monte Carlo method.

Let H be a measurable subset of $\mathbb{T}^2$. For every $j = 1,\dots,N$ let $d\mu_j$ denote the Lebesgue measure on $\mathbb{T}^2$. We define the Monte Carlo error (we also call it the *Monte Carlo discrepancy*)[3]

$$D_N^{\mathfrak{MC}} := \left\{ \int_{\mathbb{T}^2} \cdots \int_{\mathbb{T}^2} \left| N|H| - \sum_{j=1}^{N} \chi_H(u_j) \right|^2 du_1 \cdots du_N \right\}^{1/2}.$$

This term does not change if we introduce an extra translation, which allows us to apply Parseval's identity. Then by (10.15) we have

$$\left(D_N^{\mathfrak{MC}} \right)^2 = \int_{\mathbb{T}^2} \cdots \int_{\mathbb{T}^2} \int_{\mathbb{T}^2} \left| N|H| - \sum_{j=1}^{N} \chi_{H-s}(u_j) \right|^2 ds\, du_1 \cdots du_N \quad (10.16)$$

$$= \int_{\mathbb{T}^2} \cdots \int_{\mathbb{T}^2} \sum_{0 \neq k \in \mathbb{Z}^2} \left| \widehat{\chi}_H(k) \right|^2 \left| \sum_{j=1}^{N} e^{2\pi i k \cdot u_j} \right|^2 du_1 \cdots du_N$$

$$= \sum_{0 \neq k \in \mathbb{Z}^2} \left| \widehat{\chi}_H(k) \right|^2 \int_{\mathbb{T}^2} \cdots \int_{\mathbb{T}^2} \sum_{j=1}^{N} \sum_{\ell=1}^{N} e^{2\pi i k \cdot u_j} e^{-2\pi i k \cdot u_\ell}\, du_1 \cdots du_N$$

$$= \sum_{0 \neq k \in \mathbb{Z}^2} \left| \widehat{\chi}_H(k) \right|^2 \left(N + \sum_{j \neq \ell} \int_{\mathbb{T}^2} \int_{\mathbb{T}^2} e^{2\pi i k \cdot u_j} e^{-2\pi i k \cdot u_\ell}\, du_j du_\ell \right)$$

$$= N \sum_{0 \neq k \in \mathbb{Z}^2} \left| \widehat{\chi}_H(k) \right|^2 = N \left(-|H|^2 + \|\chi_H\|^2_{L^2(\mathbb{T}^2)} \right) = N \left(|H| - |H|^2 \right).$$

Observe that, up to sets of measure zero, we have $|H| - |H|^2 = 0$ if and only if $H = \varnothing$ or $H = \mathbb{T}^2$. In all other cases we have $D_N^{\mathfrak{MC}} = c\sqrt{N}$.

One would guess that suitable choices of the sequence $\{t(j)\}_{j=1}^{N}$ should improve the order of increasing $\sqrt{N}$ of the Monte Carlo discrepancy. The following theorem, proved independently by Beck [13] and Montgomery [123] (see also [28, 47] and the results of Schmidt [152, 153]) shows that the L^2 discrepancy cannot go beyond the lower bound $\sqrt[4]{N}$.

Theorem 10.7 (Beck–Montgomery) *For every convex body $C \subset \mathbb{T}^2$ having diameter less than 1 there exists a constant $c > 0$ such that for every finite set $\{t(j)\}_{j=1}^{N} \subset \mathbb{T}^2$ we have*

$$\left\{ \int_0^1 \int_{SO(2)} \int_{\mathbb{T}^2} |D_N(\varepsilon, \sigma, t)|^2\, dt d\sigma d\varepsilon \right\}^{1/2} \geq cN^{1/4}. \quad (10.17)$$

---

[3] The comparison between the Monte Carlo method and the Riemann sums associated with low-discrepancy sequences is a classical problem in the study of high-dimensional integration. See e.g. [131, Ch. 1]; also [71] and [158, Ch. 2].

Proof   Let $0 < q < 1$ and $0 < r \le 1$. By (10.15) and Lemma 10.4 we have

$$\frac{1}{r} \int_{qr}^{r} \int_{SO(2)} \int_{\mathbb{T}^2} |D_N(\varepsilon, \sigma, t)|^2 \, dt d\sigma d\varepsilon \tag{10.18}$$

$$= \sum_{m \ne 0} \left| \sum_{j=1}^{N} e^{2\pi i t(j) \cdot m} \right|^2 \frac{1}{r} \int_{qr}^{r} \int_{SO(2)} \left| \varepsilon^2 \widehat{\chi}_C(\varepsilon \sigma(m)) \right|^2 d\sigma d\varepsilon$$

$$\approx \sum_{m \ne 0} \left| \sum_{j=1}^{N} e^{2\pi i t(j) \cdot m} \right|^2 \frac{1}{r} \int_{\{qr \le |\xi| \le r\}} \left| \widehat{\chi}_C(|m| \xi) \right|^2 |\xi|^3 d\xi$$

$$\approx \sum_{m \ne 0} \left| \sum_{j=1}^{N} e^{2\pi i t(j) \cdot m} \right|^2 r^2 |m|^{-2} \int_{\{qr|m| \le |\xi| \le r|m|\}} \left| \widehat{\chi}_C(\xi) \right|^2 d\xi$$

$$\approx \sum_{m \ne 0} \left| \sum_{j=1}^{N} e^{2\pi i t(j) \cdot m} \right|^2 r^2 |m|^{-2} (1 + r |m|)^{-1},$$

where the implicit constants associated with $\approx$ do not depend on N or r. We now apply (10.18) first with $r = 1$ and second with $r = kN^{-1/2}$ (the constant k will be chosen later):

$$\int_{q}^{1} \int_{SO(2)} \int_{\mathbb{T}^2} |D_N(\varepsilon, \sigma, t)|^2 \, dt d\sigma d\varepsilon \tag{10.19}$$

$$\approx \sum_{m \ne 0} \left| \sum_{j=1}^{N} e^{2\pi i t(j) \cdot m} \right|^2 |m|^{-3}$$

$$\ge c \inf_{m \ne 0} \left\{ |m|^{-1} k^{-2} N \left(1 + kN^{-1/2} |m| \right) \right\}$$

$$\times \left\{ \sum_{m \ne 0} \left| \sum_{j=1}^{N} e^{2\pi i t(j) \cdot m} \right|^2 k^2 N^{-1} |m|^{-2} (1 + kN^{-1/2} |m|)^{-1} \right\}$$

$$\approx k^{-1} N^{1/2} \left\{ k^{-1} N^{1/2} \int_{qkN^{-1/2}}^{kN^{-1/2}} \int_{SO(2)} \int_{\mathbb{T}^2} |D_N(\varepsilon, \sigma, t)|^2 \, dt d\sigma d\varepsilon \right\}.$$

Since $qkN^{-1/2} \le \varepsilon \le kN^{-1/2}$ there exists $\delta > 0$ such that, for suitable constants q and k, we have

$$\delta \le q^2 k^2 |C| \le N\varepsilon^2 |C| \le k^2 |C| \le 1 - \delta.$$

Then[4]

$$|D_N(\varepsilon, \sigma, t)| = \left| \text{an integer} - N\varepsilon^2 |C| \right| \ge \delta$$

---

[4] This is a *trivial estimate* and the proof consists of blowing up this error.

for every σ, t and $\varepsilon \in \left[qkN^{-1/2}, kN^{-1/2} \right]$. Then

$$N^{1/2} \int_{qkN^{-1/2}}^{kN^{-1/2}} \int_{SO(2)} \int_{\mathbb{T}^2} |D_N(\varepsilon, \sigma, t)|^2 \, dt \, d\sigma \, d\varepsilon > c > 0 \,. \tag{10.20}$$

Finally, by (10.19) we have

$$\int_q^1 \int_{SO(2)} \int_{\mathbb{T}^2} |D_N(\varepsilon, \sigma, t)|^2 \, dt \, d\sigma \, d\varepsilon \geq cN^{1/2} \,.$$

$\square$

Corollary 10.8 *For every convex body $C \subset \mathbb{T}^2$ having diameter less than 1 and for every finite set $\{t(j)\}_{j=1}^N \subset \mathbb{T}^2$ there exists a (dilated, translated and rotated) copy $\widetilde{C}$ of C such that*

$$\left| D_N^{\widetilde{C}, \{t(1), \dots, t(N)\}} \right| \geq cN^{1/4} \,.$$

We show that in certain cases the dilation is not necessary (and in other cases it cannot be avoided). We first need an estimate (essentially due to Cassels [40, 123]) for the sums $\sum_{j=1}^N e^{2\pi i m \cdot t(j)}$ which appear in the Fourier coefficients of $D_N(t)$, see (10.15).

Lemma 10.9 (Cassels) *For every positive integer N let*

$$Q_N := \left\{ m = (m_1, m_2) \in \mathbb{Z}^2 : |m_1| \leq \sqrt{2N}, \, |m_2| \leq \sqrt{2N} \right\} \,. \tag{10.21}$$

Then for every finite set $\{t(j)\}_{j=1}^N \subset \mathbb{T}^2$ we have

$$\sum_{0 \neq m \in Q_N} \left| \sum_{j=1}^N e^{2\pi i m \cdot t(j)} \right|^2 \geq N^2 \,. \tag{10.22}$$

Proof We add N^2 on both sides of (10.22), which becomes

$$\sum_{|m_1| \leq \sqrt{2N}} \sum_{|m_2| \leq \sqrt{2N}} \left| \sum_{j=1}^N e^{2\pi i m \cdot t(j)} \right|^2 \geq 2N^2 \,.$$

It is enough to prove that

$$\sum_{|m_1| \leq \lfloor \sqrt{2N} \rfloor} \sum_{|m_2| \leq \lfloor \sqrt{2N} \rfloor} \left| \sum_{j=1}^N e^{2\pi i m \cdot t(j)} \right|^2 \geq N \left(\lfloor \sqrt{2N} \rfloor + 1 \right)^2 \,. \tag{10.23}$$

Let $t(\ell) = (t_1(\ell), t_2(\ell))$. Then the LHS of (10.23) is larger than

$$\sum_{|m_1|\le\left[\sqrt{2N}\right]}\sum_{|m_2|\le\left[\sqrt{2N}\right]}\left(1-\frac{|m_1|}{\left[\sqrt{2N}\right]+1}\right)\left(1-\frac{|m_2|}{\left[\sqrt{2N}\right]+1}\right)\left|\sum_{j=1}^{N}e^{2\pi i m\cdot t(j)}\right|^2$$

$$=\sum_{|m_1|\le\left[\sqrt{2N}\right]}\sum_{|m_2|\le\left[\sqrt{2N}\right]}\left(1-\frac{|m_1|}{\left[\sqrt{2N}\right]+1}\right)\left(1-\frac{|m_2|}{\left[\sqrt{2N}\right]+1}\right)$$

$$\times\sum_{j=1}^{N}\sum_{k=1}^{N}e^{2\pi i m\cdot(t(j)-t(k))}$$

$$=\sum_{j=1}^{N}\sum_{k=1}^{N}\sum_{|m_1|\le\left[\sqrt{2N}\right]}\left(1-\frac{|m_1|}{\left[\sqrt{2N}\right]+1}\right)e^{2\pi i m_1(t_1(j)-t_1(k))}$$

$$\times\sum_{|m_2|\le\left[\sqrt{2N}\right]}\left(1-\frac{|m_2|}{\left[\sqrt{2N}\right]+1}\right)e^{2\pi i m_2(t_2(j)-t_2(k))}$$

$$=\sum_{j=1}^{N}\sum_{k=1}^{N}K_{\left[\sqrt{2N}\right]}(t_1(j)-t_1(k))\,K_{\left[\sqrt{2N}\right]}(t_2(j)-t_2(k))\,,\qquad(10.24)$$

where K_M is the Fejér kernel on $\mathbb{T}$ (see (6.7)). Since $K_M(x)\ge 0$ for every x, the last term in (10.24) is not smaller than the 'diagonal':

$$\sum_{j=1}^{N}K_{\left[\sqrt{2N}\right]}(t_1(j)-t_1(j))\,K_{\left[\sqrt{2N}\right]}(t_2(j)-t_2(j))$$

$$=N\,K_{\left[\sqrt{2N}\right]}(0)\,K_{\left[\sqrt{2N}\right]}(0)=N\left(\left[\sqrt{2N}\right]+1\right)^2.$$

$\square$

We have the following result [172].

Theorem 10.10 *Let $T\subset\mathbb{T}^2$ be a triangle having sides of length less than 1. Then for every finite set $\{t(j)\}_{j=1}^{N}\subset\mathbb{T}^2$ we have*

$$\int_{\mathbb{T}^2}\int_{SO(2)}|D_N(\sigma,t)|^2\,d\sigma dt\ge c\,N^{1/2}.$$

Corollary 10.11 *Given a triangle $T\subset\mathbb{T}^2$ having sides of length less than 1, and given a finite set $\{t(j)\}_{j=1}^{N}\subset\mathbb{T}^2$, there exists a copy (translated and rotated) $\widetilde{T}$ of T such that*

$$\left|-N\,|T|+\operatorname{card}\left(\widetilde{T}\cap\{t(j)\}_{j=1}^{N}\right)\right|\ge cN^{1/4}.$$

Proof of Theorem 10.10   Applying (10.15) with $\varepsilon = 1$, Parseval's identity and Lemmas 10.6 and 10.9 we obtain

$$\int_{\mathbb{T}^2} \int_{SO(2)} |D_N(\sigma, t)|^2 \, d\sigma dt = \sum_{0 \neq m \in \mathbb{Z}^2} \left| \sum_{j=1}^{N} e^{2\pi i m \cdot t(j)} \right|^2 \int_{SO(2)} \left| \hat{\chi}_T(\sigma(m)) \right|^2 \, d\sigma$$

$$\geq c \sum_{0 \neq m \in Q_N} \left| \sum_{j=1}^{N} e^{2\pi i m \cdot t(j)} \right|^2 |m|^{-3} \geq c N^{-3/2} \sum_{0 \neq m \in Q_N} \left| \sum_{j=1}^{N} e^{2\pi i m \cdot t(j)} \right|^2 \geq c \, N^{1/2} \,,$$

where Q_N has been defined in (10.21).   □

A variant of the above argument can be used to give a different proof of Roth's theorem (see [123, Ch. 6]).

So far we have seen two different estimates from below for the L^2 discrepancy: the logarithmic estimate in Roth's theorem and the $N^{1/4}$ estimate proved in this section. The logarithmic estimate is optimal, since Theorem 8.24 gives an upper counterpart of Roth's theorem. Now we show also that Theorem 10.7 and Corollary 10.4 cannot be improved. This result is due to Beck and Chen [15], and we provide two proofs in this chapter. The first one will depend on Kendall's result (Theorem 8.20 [106], see also [35]). The second one is probabilistic in nature and will be a particular case of an L^p result. A third proof will be given in the next chapter (Remark 11.5).

Theorem 10.12 (Beck and Chen)   *Let $C \subset [-1/2, 1/2]^2$ be a convex body with diameter less than 1. Then for every positive integer N there exists a finite set $\{t(j)\}_{j=1}^{N} \subset \mathbb{T}^2$ such that*

$$\int_{SO(2)} \int_{\mathbb{T}^2} |D_N(\sigma_\theta, t)|^2 \, dt d\theta \leq c N^{1/2} \,. \tag{10.25}$$

Proof   By Lagrange's theorem (Theorem 5.23) there are four non-negative integers j_1, j_2, j_3, j_4 such that $N = j_1^2 + j_2^2 + j_3^2 + j_4^2$. Arguing as in [34] we choose $a_1, a_2, a_3, a_4 \in [0, 1)^2$ such that

$$\left(a_\ell + j_\ell^{-1} \mathbb{Z}^2 \right) \cap \left(a_k + j_k^{-1} \mathbb{Z}^2 \right) = \varnothing \tag{10.26}$$

whenever $\ell \neq k$. For each $\ell = 1, 2, 3, 4$, the set

$$A_{j_\ell^2} := \left(a_\ell + j_\ell^{-1} \mathbb{Z}^2 \right) \cap [0, 1)^2$$

contains j_ℓ^2 elements. By (10.26) the union $A_N = A_{j_1^2} \cup A_{j_2^2} \cup A_{j_3^2} \cup A_{j_4^2}$ is disjoint and therefore it contains N points. Observe that for every measurable set $H \subset \mathbb{T}^2$ we have

$$\text{card} \, (A_N \cap H) - N |H|$$

$$= \operatorname{card}\left(A_{\hat{f}_1} \cap H\right) - \hat{f}_1^2 |H| + \ldots + \operatorname{card}\left(A_{\hat{f}_4} \cap H\right) - \hat{f}_4^2 |H| \ .$$

Hence (10.25) will be a consequence of, say,

$$\int_{SO(2)} \int_{\mathbb{T}^2} \left| \operatorname{card}\left(A_{\hat{f}_1} \cap (\sigma(C) + t)\right) - \hat{f}_1^2 |C| \right|^2 dt\, d\theta \le cN^{1/2} \ .$$

In other words, we may prove the theorem assuming $N = M^2$ (a square) and considering the finite set

$$\{t(j)\}_{j=1}^N = A_N = A_{M^2} = \left\{a + \left(\frac{p}{M}, \frac{q}{M}\right)\right\}_{p,q \in \mathbb{Z}} \cap [0, 1)^2 \ ,$$

where $a \in \left[0, M^{-1}\right)^2$ is given. Then

$$\int_{SO(2)} \int_{\mathbb{T}^2} \left| \operatorname{card}\left(A_{M^2} \cap (\sigma(C) + t)\right) - M^2 |C| \right|^2 dt\, d\sigma$$

$$= \int_{SO(2)} \int_{\mathbb{T}^2} \left| \operatorname{card}\left(A_{M^2} \cap (\sigma(C) + t + a)\right) - M^2 |C| \right|^2 dt\, d\sigma$$

$$= \int_{SO(2)} \int_{\mathbb{T}^2} \left| \operatorname{card}\left(\left\{\left(\frac{p}{M}, \frac{q}{M}\right)\right\}_{p,q=0}^{M-1} \cap (\sigma(C) + t)\right) - M^2 |C| \right|^2 dt\, d\sigma$$

$$= \int_{SO(2)} \int_{[0,1)^2} \left| \operatorname{card}\left(\{(p, q)\}_{p,q \in \mathbb{Z}} \cap (\sigma(MC) + Mt)\right) - M^2 |C| \right|^2 dt\, d\sigma$$

$$= M^{-2} \int_{SO(2)} \int_{[0,M)^2} \left| \operatorname{card}\left(\mathbb{Z}^2 \cap (M\sigma(C) + v)\right) - M^2 |C| \right|^2 dv\, d\sigma$$

$$= \int_{SO(2)} \int_{\mathbb{T}^2} \left| \operatorname{card}\left(\mathbb{Z}^2 \cap (M\sigma(C) + v)\right) - M^2 |C| \right|^2 dv\, d\sigma \ , \qquad (10.27)$$

since the function

$$v \longmapsto \operatorname{card}\left(\mathbb{Z}^2 \cap (M\sigma(C) + v)\right) - M^2 |C|$$

is periodic and the square $[0, M)^2$ consists of M^2 disjoint copies of $[0, 1)^2$. In this way we have changed our point of view slightly. Now M is no longer the (square root of the) number of points, but rather a (discrete) dilation parameter, while (10.27) is the lattice point problem studied in Theorem 8.26. Then

$$\int_{SO(2)} \int_{\mathbb{T}^2} \left| \operatorname{card}\left(A_{M^2} \cap (\sigma(C) + t)\right) - M^2 |C| \right|^2 dt\, d\sigma \le cM = cN^{1/2} \ .$$

$\square$

The second proof of Theorem 10.12 comes from the case $p = 2$ in the next theorem, proved by Chen [22, 44]. Let M be a large positive integer and let

$N = M^2$. Let

$$\{t(j)\}_{j=1}^N := \frac{1}{M} \mathbb{Z}^2 \cap \left[-\frac{1}{2}, \frac{1}{2}\right)^2 \tag{10.28}$$

be the restriction of the shrunk lattice $\frac{1}{M}\mathbb{Z}^2$ to the unit square. We replace every point $t(j)$ with a random point u_j in the small square

$$S_j = t(j) + \left[-\frac{1}{2M}, \frac{1}{2M}\right)^2 \tag{10.29}$$

centred at $t(j)$ and having side length $1/M$ (in statistics this choice may be termed *jittered discrepancy*, see e.g. [16, 110]).

Theorem 10.13 (Chen)   *Let $C \subset [0,1)^2$ be a convex body. Let M be a large positive integer and let $N = M^2$. Let $S_1, \ldots, S_N$ be the N small squares with area $1/N$ in (10.29). Then for every $p < +\infty$ we have (see (10.14))*

$$\left\{ N^N \int_{S_N} \cdots \int_{S_1} \left|D_N^{C,\{u_1,\ldots,u_N\}}\right|^p \, du_1 \cdots du_N \right\}^{1/p} \leq c_p \, N^{1/4},$$

where c_p is independent of N and, for every j, du_j is the Lebesgue measure.

Proof   By Corollary 6.6 we may assume that p is an even integer. For every S_j we have

$$\int_{S_j} \chi_{C \cap S_j}(u_j) \, du_j = |C \cap S_j| . \tag{10.30}$$

Since C is convex, there are $M^\#$ ($\approx M$) small squares which intersect the boundary of C. Up to a reordering, we may write the set of these small squares as $\mathcal{A} = \{S_k\}_{k=1}^{M^\#}$. Observe that for every $S \notin \mathcal{A}$ and every $u_S \in S$ we have

$$\chi_{C \cap S}(u_S) - N|C \cap S| = 0 . \tag{10.31}$$

For every $j = 1, \ldots, N$ let $u_j \in S_j$. Then (10.31) implies

$$D_N^{C,\{u_1,\ldots,u_N\}} = -N|C| + \sum_{j=1}^N \chi_C(u_j) = -N \sum_{j=1}^N |C \cap S_j| + \sum_{j=1}^N \chi_{C \cap S_j}(u_j)$$

$$= \sum_{S \in \mathcal{A}} (\chi_{C \cap S}(u_S) - N|C \cap S|) .$$

Hence

$$\left(D_N^{C,\{u_1,\ldots,u_N\}}\right)^p$$

$$= \sum_{S_1,\ldots,S_p \in \mathcal{A}} \{N|C \cap S_1| - \chi_{B \cap S_1}(u_{S_1})\} \cdots \{N|C \cap S_p| - \chi_{B \cap S_p}(u_{S_p})\} .$$

We have

$$N^N \int_{S_N} \cdots \int_{S_1} \left(D_N^{C,\{u_1,\ldots,u_N\}} \right)^p du_1 \cdots du_N \tag{10.32}$$

$$= N^N \int_{S_N} \cdots \int_{S_1} \sum_{S_1,\ldots,S_p \in \mathcal{A}}$$

$$\times \left(\chi_{C \cap S_1}(u_{S_1}) - N \,|C \cap S_1| \right) \cdots \left(\chi_{C \cap S_p}(u_{S_p}) - N \,|C \cap S_p| \right) du_1 \cdots du_N$$

$$= N^{M^\#} \sum_{S_1,\ldots,S_p \in \mathcal{A}} \int_{S_{M^\#}} \cdots \int_{S_1}$$

$$\times \left(\chi_{C \cap S_1}(u_{S_1}) - N \,|C \cap S_1| \right) \cdots \left(\chi_{C \cap S_p}(u_{S_p}) - N \,|C \cap S_p| \right) du_1 \cdots du_{M^\#}.$$

By (10.30) the term in (10.32) is zero whenever at least one of the terms

$$\chi_{C \cap S_i}(u_{S_i}) - N \,|C \cap S_i|$$

appears exactly once in the above product. Then the non-zero contributions to the sum $\sum_{S_1,\ldots,S_p \in \mathcal{A}}$ in (10.32) may come only when each term appears at least twice. For simplicity, let us first think that the small squares $S_1, S_2, \ldots, S_p \in \mathcal{A}$ are subdivided into $p/2$ pairs, which can be chosen in

$$\binom{M^\#}{p/2} = \frac{M^\# \left(M^\# - 1 \right) \cdots \left(M^\# - p/2 + 1 \right)}{(p/2)!} \leq c_p \, M^{p/2}$$

ways. In general, the small squares are grouped into sets of two, three, four, ... Then we have to choose $q \leq p/2$ sets of two, three, four, ... This can be done in at most $c_p' \, M^{p/2}$ ways. Observe that we have

$$\left| \chi_{C \cap S_j}(u_j) - N \,|C \cap S_j| \right| \leq 1$$

for every j and every $u_j \in S_j$. Then

$$N^N \int_{S_N} \cdots \int_{S_1} \left(D_N^{C,\{u_1,\ldots,u_N\}} \right)^p du_1 \cdots du_N \leq c_p''' \, M^{p/2} = c_p''' \, N^{p/4}.$$

$$\square$$

For the case $p = \infty$ a different argument due to Beck gives the upper estimate $cN^{1/4} \log^{1/2} N$ [12, 14, 50].

Exercises

1) Let $Q = [-1/2, 1/2]^2$ and let $1 < p < \infty$. Prove the existence of a positive constant c_p such that

$$\left\|\widehat{\chi_Q}(\rho \cdot)\right\|_{L^p([0,2\pi))} \geq c_p \, \rho^{-3/2-1/(2p)}$$

for every real $\rho \geq 2$.

2) Let $Q = [-1/2, 1/2]^2$ and let $1 < p < +\infty$. Prove the existence of a sequence $\rho_k \to +\infty$ and a positive constant c_p such that

$$\left\|\widehat{\chi_Q}(\rho_k \cdot)\right\|_{L^p([0,2\pi))} \geq c_p \, \rho_k^{-1-1/p} \ .$$

3) Use the argument in the proof of Lemma 10.6 to give another proof of Lemma 8.22.

4) Prove that the Fourier transform of the characteristic function $\chi_1(t)$ of the unit disc $B(0, 1) \subset \mathbb{R}^2$, centred at the origin, admits a diverging sequence of zeros.

5) Let T be a triangle in $\mathbb{R}^2$ and let $1 < p < \infty$. Prove the existence of a constant $c_p > 0$ such that, for large R,

$$\left\{ \int_{SO(2)} \int_{\mathbb{T}^2} \left| \text{card} \left(\mathbb{Z}^2 \cap (R\sigma_\theta(T) + t) \right) - R^2 |T| \right|^p \, dt d\theta \right\}^{1/p} \geq c \, R^{1-1/p} \ ,$$

where σ_θ denotes rotation by the angle θ.

6) Let $Q = [-1/2, 1/2]^2$. For $0 \leq x < 1/4$ and $0 \leq \theta < 2\pi$ let $\pi(x, \theta)$ be the half-plane with slope θ and distance x from the origin. Let $A_{\theta,t} = Q \cap \pi(t, \theta)$. Prove the existence of a sequence $N \to +\infty$ of positive integers such that there exists a distribution $\{u_j\}_{j=1}^{N} \subset Q$ satisfying

$$\int_{\mathbb{T}^2} \int_0^{2\pi} \left| N \, |A_{\theta,t}| - \sum_{j=1}^{N} \chi_{A_{\theta,t}} (u_j) \right| \, d\theta dt \leq c \, \log^2(N) \ .$$

7) Let $\{u_j\}_{j=1}^{N} \subset [-1/2, 1/2]^2$. Prove the existence of a convex body $C \subset [-1/2, 1/2]^2$ and a positive constant c such that

$$\left| D_N^{C, \{u_1, \ldots, u_N\}} \right| \geq c \, N^{1/3} \ .$$

11

Discrepancy in high dimension and Bessel functions

In this chapter we investigate the discrepancy associated with the translates of a d-dimensional ball rB in $\mathbb{T}^d$ (here B is the unit ball centred at the origin, and $0 < r < 1/2$). We denote by χ_1 the characteristic function of B. In Lemma 8.14 we used a geometric argument to prove an upper bound for $\widehat{\chi_1}(\xi)$ when $d = 2$. Here we shall use complex analysis to obtain, for every d, an asymptotic estimate of $\widehat{\chi_1}(\xi)$, as $|\xi| \to +\infty$. We start by recalling a probably well-known result.

Lemma 11.1 *The d-dimensional unit ball $B = \left\{ t \in \mathbb{R}^d : |t| \leq 1 \right\}$ has volume*

$$\omega_d = \frac{\pi^{d/2}}{\Gamma\left(\frac{d}{2} + 1\right)} ,$$

where Γ is the gamma function (see (5.18)).

Proof By the d-dimensional integral formula in spherical coordinates there exists $H_d > 0$ such that for every continuous radial integrable function[1] f on $\mathbb{R}^d$ we have

$$\int_{\mathbb{R}^d} f(t) \, dt = H_d \int_0^{+\infty} f_o(r) \, r^{d-1} \, dr , \tag{11.1}$$

where f_o is defined by $f_o(|t|) = f(t)$. We are going to compute H_d. On the one hand, (6.25) gives

$$\int_{\mathbb{R}^d} e^{-|t|^2} \, dt = \int_{\mathbb{R}} e^{-t_1^2} \, dt_1 \cdots \int_{\mathbb{R}} e^{-t_d^2} \, dt_d = \pi^{d/2} .$$

On the other hand, (11.1) gives

$$\int_{\mathbb{R}^d} e^{-|t|^2} \, dt = H_d \int_0^{+\infty} e^{-r^2} \, r^{d-1} \, dr = \frac{1}{2} H_d \int_0^{+\infty} e^{-x} \, x^{(d-2)/2} \, dx$$

---

[1] We say that a function $f(t)$ on $\mathbb{R}^d$ is radial if $f(t) = f(u)$ whenever $|t| = |u|$. It is easy to prove that the Fourier transform of a radial integrable function is a radial function.

$$= \frac{1}{2} H_d \, \Gamma\left(\frac{d}{2}\right) .$$

Then

$$H_d = \frac{2\pi^{d/2}}{\Gamma(d/2)} .$$

Therefore, (5.19) implies

$$\omega_d = H_d \int_0^1 r^{d-1} \, dr = \frac{1}{d} \frac{2\pi^{d/2}}{\Gamma\left(\frac{d}{2}\right)} = \frac{\pi^{d/2}}{\Gamma\left(\frac{d}{2} + 1\right)} .$$

□

Observe that $\omega_d \to 0$ as $d \to +\infty$.

Let $\chi_1(t)$ be the characteristic function of the d-dimensional unit ball centred at the origin. If we argue as in the proof of Theorem 8.13 we obtain

$$\widehat{\chi_1}(\xi) = \int_{-1}^1 \omega_{d-1}\left(\sqrt{1-s^2}\right)^{d-1} e^{-2\pi i |\xi| s} \, ds \qquad (11.2)$$

$$= \frac{\pi^{(d-1)/2}}{\Gamma\left(\frac{d+1}{2}\right)} \int_{-1}^1 \left(1-s^2\right)^{(d-1)/2} e^{-2\pi i |\xi| s} \, ds .$$

The Fourier transform $\widehat{\chi_1}(\xi)$ can be expressed in terms of Bessel functions, a field of great relevance and independent interest. See, for example, [81, 117, 164, 178].

11.1 Bessel functions

For $v > -1/2$ and $x > 0$, let

$$J_v(x) := \frac{(x/2)^v}{\Gamma(v+1/2)\sqrt{\pi}} \int_{-1}^1 \left(1-s^2\right)^{v-1/2} e^{isx} \, ds \qquad (11.3)$$

be the *Bessel function*[2] (of the first kind) of order v. See, for example, [164, p. 155]. Then (11.2) yields

$$\widehat{\chi_1}(\xi) = \rho^{-d/2} J_{d/2}(2\pi |\xi|) . \qquad (11.4)$$

---

[2] (11.3) is known as the Poisson representation of Bessel functions. Another way to introduce Bessel functions is to consider the wave equation in $\mathbb{R}^2$, write the Laplacian in polar coordinates and separate variables [81]. This leads to the identity

$$J_v(x) = \sum_{k=0}^{+\infty} \frac{(-1)^k}{k!\,\Gamma(v+k+1)} \left(\frac{x}{2}\right)^{2k+v} .$$

If rB is the ball centred at the origin and having radius r, then we have the identity

$$\widehat{\chi_{rB}}(\xi) = r^{d/2} |\xi|^{-d/2} J_{d/2}(2\pi r |\xi|) . \tag{11.5}$$

A variant of the above computation shows that, for every radial $f \in L^1(\mathbb{R}^d)$ and f_o defined by $f(t) = f_o(|t|)$ for almost every $t \in \mathbb{R}^d$, we have the following identity:

$$\widehat{f}(\xi) = 2\pi |\xi|^{-(d-2)/2} \int_0^{+\infty} f_o(x) \, J_{(d-2)/2}(2\pi x |\xi|) \, x^{d/2} \, dx$$

(see [164, p. 155]).

(11.4) shows that here we only need Bessel functions of integer order n or of half-integer order $n + 1/2$.

For $J_n(x)$ we have the identity

$$J_n(x) = \frac{1}{2\pi} \int_{-\pi}^{\pi} e^{ix \sin \theta - in\theta} d\theta ,$$

while the Bessel functions $J_{n+1/2}(x)$ can be written in terms of trigonometric functions:

$$J_{1/2}(x) = \sqrt{\frac{2}{\pi x}} \sin x , \quad J_{3/2}(x) = \sqrt{\frac{2}{\pi x}} \left(\frac{\sin x}{x} - \cos x \right) , \dots$$

$$J_{n+1/2}(x) = (-1)^n \sqrt{\frac{2}{\pi}} x^{n+1/2} \left(\frac{1}{x} \frac{d}{dx} \right)^n \left(\frac{\sin x}{x} \right)$$

(see e.g. [117, Ch. 5] and [178]).

We shall estimate $J_\nu(x)$ for large x.

Theorem 11.2   *For every $\nu > -1/2$ there exists a positive constant c_ν such that, for $x \geq 1$,*

$$J_\nu(x) = \sqrt{\frac{2}{\pi x}} \cos \left(x - \frac{\nu \pi}{2} - \frac{\pi}{4} \right) + E_\nu(x) , \tag{11.6}$$

where $|E_\nu(x)| \leq c_\nu \, x^{-3/2}$.

Proof   We consider the region of the complex plane obtained by deleting the rays $(-\infty, -1]$ and $[1, +\infty)$. We then choose the branch of $\left(1 - z^2\right)^{\nu-1/2}$ which is positive on $(-1, 1)$. We compute the integral in (11.3) via Cauchy's integral theorem. We integrate

$$f(z) = \left(1 - z^2\right)^{\nu-1/2} e^{izx}$$

over the contour in the figure below, where R and ε are respectively a large positive number a small positive number

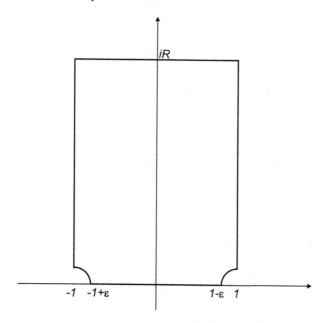

Let $\varepsilon \to 0$ and $R \to +\infty$. Then the two integrals on the small arcs vanish, as well as the integral on the upper side $\{t + iR : -1 \le t \le 1\}$. In this way we obtain

$$\int_{-1}^{1} \left(1 - t^2\right)^{\nu-1/2} e^{itx} \, dt \tag{11.7}$$

$$= \int_{0}^{+\infty} \left(1 - (-1 + iy)^2\right)^{\nu-1/2} e^{i(-1+iy)x} i \, dy$$

$$- \int_{0}^{+\infty} \left(1 - (1 + iy)^2\right)^{\nu-1/2} e^{i(1+iy)x} i \, dy$$

$$= ie^{-ix} \int_{0}^{+\infty} \left(y^2 + 2iy\right)^{\nu-1/2} e^{-yx} \, dy - ie^{ix} \int_{0}^{+\infty} \left(y^2 - 2iy\right)^{\nu-1/2} e^{-yx} \, dy .$$

We have

$$\left(y^2 \pm 2iy\right)^{\nu-1/2} = y^{\nu-1/2} \left(\pm 2i\right)^{\nu-1/2} + A(y) ,$$

where

$$|A(y)| \le c_\nu \begin{cases} y^{\nu+1/2} & \text{if } 0 < y \le 1, \\ y^{2\nu-1} & \text{if } y > 1. \end{cases}$$

Then, as $x \to +\infty$,

$$\int_0^{+\infty} \left(y^2 \pm 2iy\right)^{\nu-1/2} e^{-yx} \, dy$$

$$= (\pm 2i)^{\nu-1/2} \int_0^{+\infty} y^{\nu-1/2} e^{-yx} \, dy$$

$$+ O\left(\int_0^1 y^{\nu+1/2} e^{-yx} \, dy\right) + O\left(\int_1^{+\infty} y^{2\nu-1} e^{-yx} \, dy\right)$$

$$= (\pm 2i)^{\nu-1/2} x^{-\nu-1/2} \Gamma\left(\nu + \frac{1}{2}\right)$$

$$+ O\left(x^{-\nu-3/2} \int_0^x s^{\nu+1/2} e^{-s} \, ds\right) + O\left(x^{-2\nu} \int_x^{+\infty} s^{2\nu-1} e^{-s} \, ds\right)$$

$$= (\pm 2i)^{\nu-1/2} x^{-\nu-1/2} \Gamma\left(\nu + \frac{1}{2}\right) + O\left(x^{-\nu-3/2}\right) .$$

Then (11.3) and (11.7) imply

$$J_\nu(x) = \frac{(x/2)^\nu}{\Gamma(\nu + 1/2)\sqrt{\pi}} \int_{-1}^1 \left(1 - t^2\right)^{\nu-1/2} e^{itx} \, dt$$

$$= \frac{(x/2)^\nu}{\Gamma(\nu + 1/2)\sqrt{\pi}} x^{-\nu-1/2}$$

$$\times \left((2i)^{\nu-1/2} \Gamma\left(\nu + \frac{1}{2}\right) ie^{-ix} - (-2i)^{\nu-1/2} \Gamma\left(\nu + \frac{1}{2}\right) ie^{ix}\right) + O(x^{-3/2})$$

$$= \frac{ix^{-1/2}}{\sqrt{2\pi}} \left(i^{\nu-1/2} e^{-ix} - (-i)^{\nu-1/2} e^{ix}\right) + O(x^{-3/2})$$

$$= \frac{ix^{-1/2}}{\sqrt{2\pi}} \left(e^{i(\nu-1/2)\pi/2} e^{-ix} - e^{-i(\nu-1/2)\pi/2} e^{ix}\right) + O(x^{-3/2})$$

$$= \sqrt{\frac{2}{\pi x}} \cos\left(x - \nu\frac{\pi}{2} - \frac{\pi}{4}\right) + O(x^{-3/2}) .$$

$\square$

In the following figure we compare the graph of $J_1(x)$ with its approximation

$$\sqrt{\frac{2}{\pi x}} \cos\left(x - \frac{3}{4}\pi\right)$$

(the dashed curve):

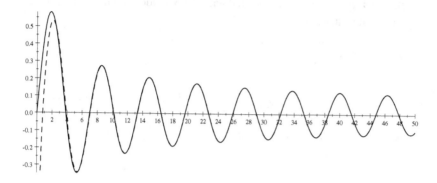

11.2 Deterministic and probabilistic discrepancies

Let M be a large positive integer and let $N = M^d$. As in (10.28) let

$$\{t(j)\}_{j=1}^{N} = \frac{1}{M}\mathbb{Z}^d \cap \left[-\frac{1}{2},\frac{1}{2}\right)^d$$

be the restriction of $M^{-1}\mathbb{Z}^d$ to the unit cube. For every $j = 1,\ldots,N$, we consider, as in Theorem 10.13, a random point inside each small cube

$$S_j = t(j) + \left[-\frac{1}{2M},\frac{1}{2M}\right)^d$$

centred at $t(j)$ and with side length $1/M$. Let

$$\lambda(t) := N\chi_{[-1/(2M),1/(2M))^d}(t) \tag{11.8}$$

be the (normalized) characteristic function of $[-1/(2M),1/(2M))^d$. For every $j = 1,\ldots,N$ let

$$\lambda_j(t) = \lambda(t - t(j)) . \tag{11.9}$$

Then $\lambda_j(t)$ is supported on S_j and $\lambda_j(t)\,dt$ is the (normalized) Lebesgue measure on S_j.

We are going to compare the *grid discrepancy* associated with the finite sequence $\{t(j)\}_{j=1}^{N}$, that is, the square root of

$$D_{\text{grid}}^2(N) := \int_{\mathbb{T}^d} \left| -N\,|rB| + \sum_{j=1}^{N}\chi_{rB-t}(t(j)) \right|^2 dt$$

and the *jittered discrepancy* associated with the N random points in the squares S_j, that is, the square root of

$$\mathcal{D}^2_{\text{jittered}}(N) := \int_{\mathbb{T}^d} \cdots \int_{\mathbb{T}^d} \int_{\mathbb{T}^d} \tag{11.10}$$

$$\times \left| -N|rB| + \sum_{j=1}^{N} \chi_{rB-t}(u_j) \right|^2 dt\lambda_1(u_1) du_1 \cdots \lambda_N(u_N) du_N \, .$$

These two choices of points are represented in the figures below:

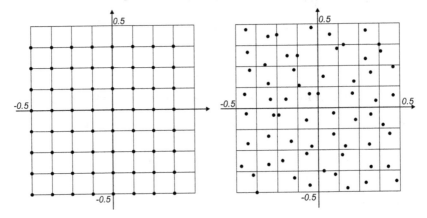

For every $k = (k_1, \ldots, k_d) \in \mathbb{Z}^d$ we have

$$\widehat{\lambda}(k) = N \prod_{s=1}^{d} \frac{\sin(\pi k_s / M)}{\pi k_s} \tag{11.11}$$

with obvious modifications when $k_s = 0$ for some $s = 1, \ldots, d$. Moreover

$$\widehat{\lambda}_j(k) = e^{2\pi i k \cdot t(j)} \widehat{\lambda}(k) \tag{11.12}$$

for every $j = 1, \ldots, N$ and $k \in \mathbb{Z}^d$.

Theorem 11.3   *Let rB be the ball centred at the origin, with radius $r < 1/2$. The jittered discrepancy satisfies the following identity:*

$$\mathcal{D}^2_{\text{jittered}}(N) = N \left(\|\chi_{rB}\|^2_{L^2(\mathbb{T}^d)} - \|\chi_{rB} * \lambda\|^2_{L^2(\mathbb{T}^d)} \right) . \tag{11.13}$$

Proof   We argue as in (10.16). By Parseval's identity and (11.12) we have

$$\mathcal{D}^2_{\text{jittered}}(N)$$

$$= \int_{\mathbb{T}^d} \cdots \int_{\mathbb{T}^d} \sum_{0 \neq k \in \mathbb{Z}^d} |\widehat{\chi}_{rB}(k)|^2 \left| \sum_{j=1}^{N} e^{2\pi i k \cdot u(j)} \right|^2 \lambda_1(u_1) du_1 \cdots \lambda_N(u_N) du_N$$

$$= \sum_{0 \neq k \in \mathbb{Z}^2} |\widehat{\chi}_{rB}(k)|^2 \sum_{j=1}^{N} \sum_{\ell=1}^{N} \int_{\mathbb{T}^d} \int_{\mathbb{T}^d} e^{2\pi i k \cdot u_j} e^{-2\pi i k \cdot u_\ell} \lambda_j(u_j) du_j \lambda_\ell(u_\ell) du_\ell$$

$$= \sum_{0 \neq k \in \mathbb{Z}^2} |\widehat{\chi}_{rB}(k)|^2 \left(N + \sum_{j \neq \ell} \int_{\mathbb{T}^d} \int_{\mathbb{T}^d} e^{2\pi i k \cdot u_j} e^{-2\pi i k \cdot u_\ell} \lambda_j(u_j) du_j \lambda_\ell(u_\ell) du_\ell \right)$$

$$= \sum_{0 \neq k \in \mathbb{Z}^2} |\widehat{\chi}_{rB}(k)|^2 \left(N + \sum_{j \neq \ell} e^{2\pi i k \cdot t(j)} e^{-2\pi i k \cdot t(\ell)} \right.$$

$$\times \int_{\mathbb{T}^d} \int_{\mathbb{T}^d} e^{2\pi i k \cdot u_j} e^{-2\pi i k \cdot u_\ell} \lambda(u_j) du_j \lambda(u_\ell) du_\ell \bigg)$$

$$= \sum_{0 \neq k \in \mathbb{Z}^2} |\widehat{\chi}_{rB}(k)|^2 \left(N + \sum_{j \neq \ell} |\widehat{\lambda}(k)|^2 \, e^{2\pi i k \cdot t(j)} e^{-2\pi i k \cdot t(\ell)} \right).$$

Hence

$$\mathcal{D}^2(N) = \sum_{0 \neq k \in \mathbb{Z}^2} |\widehat{\chi}_{rB}(k)|^2 \left(N + |\widehat{\lambda}(k)|^2 \left(-N + \left| \sum_{j=1}^{N} e^{2\pi i k \cdot t(j)} \right|^2 \right) \right) \qquad (11.14)$$

$$= N \sum_{k \in \mathbb{Z}^2} |\widehat{\chi}_{rB}(k)|^2 \left(1 - |\widehat{\lambda}(k)|^2 \right) + \sum_{0 \neq k \in \mathbb{Z}^2} |\widehat{\chi}_{rB}(k)|^2 |\widehat{\lambda}(k)|^2 \left| \sum_{j=1}^{N} e^{2\pi i k \cdot t(j)} \right|^2.$$

Observe that

$$\sum_{j=1}^{N} e^{2\pi i k \cdot t(j)} = \begin{cases} N & \text{if } k \in M\mathbb{Z}^d, \\ 0 & \text{otherwise.} \end{cases} \qquad (11.15)$$

Indeed, by (10.28) we have

$$\sum_{j=1}^{N} e^{2\pi i k \cdot t(j)} = \prod_{s=1}^{d} \left(\sum_{j=0}^{M-1} e^{2\pi i k_s j/M} \right),$$

where $k = (k_1, \ldots, k_d)$. If $k \notin M\mathbb{Z}^d$ there is $k_s \notin M\mathbb{Z}$. Then

$$\sum_{j=0}^{M-1} e^{2\pi i k_s j/M} = \sum_{j=0}^{M-1} \left(e^{2\pi i k_s/M} \right)^j = 0$$

and (11.15) is proved. Then (11.11) and (11.15) imply

$$\widehat{\lambda}(k) \sum_{j=1}^{N} e^{2\pi i k \cdot t(j)} = 0$$

for every $k \neq 0$. Hence

$$\mathcal{D}^2_{\text{jittered}}(N) = N \sum_{k \in \mathbb{Z}^2} |\widehat{\chi_{rB}}(k)|^2 \left(1 - |\widehat{\lambda}(k)|^2\right) = N \left(\|\chi_{rB}\|^2_{L^2(\mathbb{T}^d)} - \|\chi_{rB} * \lambda\|^2_{L^2(\mathbb{T}^d)}\right).$$

$\square$

Lemma 11.4 *For every dimension d we have*

$$\mathcal{D}^2_{\text{jittered}}(N) < M^{d-1} \frac{\pi^{d/2} d^{3/2} r^{d-1}}{2\Gamma\left(\frac{d}{2}+1\right)}.$$

Proof For every $t \in \mathbb{T}^d$ we have

$$(\chi_{rB} * \lambda)(t) = N \left|rB \cap \left(\left[-\frac{1}{2M}, \frac{1}{2M}\right]^d + t\right)\right|. \tag{11.16}$$

Observe that the support of the function λ is small and we have

$$(\chi_{rB} * \lambda)(t) = \chi_{rB}(t)$$

for every

$$t \notin \left\{x \in \mathbb{T}^2 : \min_{y \in \partial(rB)} |x - y| \leq \frac{\sqrt{d}}{2M}\right\}.$$

Hence (11.13) and Lemma 11.1 imply, for large M,

$$\mathcal{D}^2_{\text{jittered}}(N) \leq N |rB| - N \int_{|t| < r - \sqrt{d}/(2M)} dt$$

$$= N \left(r^d - \left(r - \frac{\sqrt{d}}{2M}\right)^d\right) \frac{\pi^{d/2}}{\Gamma\left(\frac{d}{2}+1\right)}$$

$$\leq N \frac{\pi^{d/2}}{\Gamma\left(\frac{d}{2}+1\right)} dr^{d-1}\left(r - \left(r - \frac{\sqrt{d}}{2M}\right)\right) \leq M^{d-1} \frac{d^{3/2} r^{d-1} \pi^{d/2}}{2\Gamma\left(\frac{d}{2}+1\right)}.$$

$\square$

Remark 11.5 The argument in the previous proof shows that for large M we have

$$\left|\left\{t \in \mathbb{T}^d : (\chi_{rB} * \lambda)(t) \neq \chi_{rB}(t)\right\}\right| \leq c_d/M = c_d N^{-1/d}.$$

Hence $\mathcal{D}^2_{\text{jittered}}(N) \leq c N^{1-1/d}$. This gives another proof of Theorem 10.12.

Theorem 11.6 *Let the dimension d be large and let $d \not\equiv 1 \pmod 4$. Then, for every sufficiently large positive integer M, we have*

$$\mathcal{D}^2_{\text{jittered}}\left(M^d\right) \leq D^2_{\text{grid}}\left(M^d\right). \tag{11.17}$$

Proof Let $N = M^d$. As in the proof of Theorem 10.12, we reduce to the argument in Theorem 8.26. Namely, (10.15), Parseval's equality, (11.15) and (11.5) give

$$D_{\text{grid}}^2(N) = \int_{\mathbb{T}^d} \left| -N |rB| + \sum_{j=1}^N \chi_{rB-t}(t(j)) \right|^2 dt \tag{11.18}$$

$$= \sum_{k \neq 0} \left| \sum_{j=1}^N e^{2\pi i k \cdot t(j)} \right|^2 \left| \widehat{\chi_{rB}}(k) \right|^2 = N^2 \sum_{h \neq 0} \left| \widehat{\chi_{rB}}(Mh) \right|^2$$

$$= N^2 \sum_{h \neq 0} r^d |Mh|^{-d} J_{d/2}^2(2\pi rM |h|) = Nr^d \sum_{h \neq 0} |h|^{-d} J_{d/2}^2(2\pi rM |h|) .$$

By (11.6) we have

$$J_{d/2}(2\pi rM |h|) \tag{11.19}$$

$$= \frac{1}{\pi r^{1/2} M^{1/2} |h|^{1/2}} \cos\left(2\pi rM |h| - (d+1)\frac{\pi}{4}\right) + O_d\left(\frac{1}{(rM |h|)^{3/2}}\right) .$$

Assume

$$\left| \cos\left(2\pi rM - (d+1)\frac{\pi}{4}\right) \right| \geq \frac{1}{5} . \tag{11.20}$$

Then (11.18) and (11.19) imply

$$D_{\text{grid}}^2(N) \geq Nr^d \sum_{|h|=1} J_{d/2}^2(2\pi rM |h|) \geq M^{d-1} \frac{dr^{d-1}}{1000} .$$

Of course (11.20) may fail. That is, $2\pi rM - (d+1)\pi/4$ can be close to $\pi\mathbb{Z} + \pi/2$. Then $4\pi rM - (d+1)\pi/2$ is close to $2\pi\mathbb{Z} + \pi$. Hence $4\pi rM - (d+1)\pi/4$ is close to $(2\mathbb{Z}+1)\pi + (d+1)\pi/4$, which is away from $\pi\mathbb{Z} + \pi/2$ because $d \not\equiv 1 \pmod{4}$. Then we can deduce that

$$\min\left(\left|\cos\left(2\pi rM - (d+1)\frac{\pi}{4}\right)\right|, \left|\cos\left(4\pi rM - (d+1)\frac{\pi}{4}\right)\right|\right) \geq \frac{1}{10} . \tag{11.21}$$

Then, for every large M,

$$D_{\text{grid}}^2(N) \geq Nr^d \sum_{|h|=1, |h|=2} |h|^{-d} J_{d/2}^2(2\pi rM |h|) \geq M^{d-1} \frac{dr^{d-1}}{1000} .$$

Since for large d we have

$$\frac{dr^{d-1}}{1000} > \frac{dr^{d-1}\pi^{d/2}}{2\Gamma\left(\frac{d}{2}+1\right)} ,$$

the previous lemma implies (11.17). $\square$

Remark 11.7   The inequality (11.17) is not true for every $d \not\equiv 1 \pmod 4$. A cumbersome computation [51] shows that for $d = 2$ we have

$$\mathcal{D}^2_{\text{jittered}}\left(M^2\right) > D^2_{\text{grid}}\left(M^2\right) .$$

11.3 The case $d = 5, 9, 13, \ldots$

The case $1 < d \equiv 1 \pmod 4$ is different. As a preliminary step, we prove a simultaneous approximation result (see [90, Ch. XI] and [138]), which is related to Theorem 5.3.

Lemma 11.8   *Let $\alpha_1, \alpha_2, \ldots, \alpha_n \in \mathbb{R}$ and let H be a positive integer. Then there are $p_1, p_2, \ldots, p_n \in \mathbb{Z}$ and an integer $\widetilde{M}$ such that*

$$H \le \widetilde{M} \le H^{n+1} , \qquad \left|p_j - \alpha_j \widetilde{M}\right| < 1/H \tag{11.22}$$

for every $j = 1, \ldots, n$.

We first show why this result is related to the case $d \equiv 1 \pmod 4$. Indeed, let $d = 4\ell + 1$, then (11.18) gives

$$D^2_{\text{grid}}(N) = N r^d \sum_{h \ne 0} |h|^{-d} J^2_{d/2}(2\pi r M |h|) ,$$

while (11.6) yields

$$J^2_{d/2}(2\pi r M |h|) \sim \frac{1}{\pi^2 r M |h|} \cos^2\left(2\pi r M |h| - \frac{4\ell + 1}{2}\frac{\pi}{2} - \frac{\pi}{4}\right)$$

$$= \frac{1}{\pi^2 r M |h|} \sin^2(2\pi r M |h|) .$$

Then we can use (11.22) to obtain an integer $\widetilde{M}$ and a useful upper bound for each term $\sin^2\left(2\pi r \widetilde{M} |h|\right)$, where h is any integer number contained in a suitably large ball centred at the origin.

Proof of Lemma 11.8   We may assume that $0 < \alpha_j < 1$ for every $j = 1, \ldots, n$. We subdivide the cube $[0, 1)^n$ as the disjoint union

$$[0, 1)^n = \bigcup_{k=1}^{H^n} Q_k ,$$

where the cubes Q_k have sides parallel to the axes and length H^{-1}. For every integer $0 \le m \le H^{n+1}$ we consider the point

$$(\{m\alpha_1\}, \{m\alpha_2\}, \ldots, \{m\alpha_n\}) \in [0, 1)^n ,$$

where $\{m\alpha_k\}$ is the fractional part of $m\alpha_k$. Then there exists a cube Q_{k_0} which contains at least $H + 1$ of the above $H^{n+1} + 1$ points. Hence there are integers

$$0 \le m_1 < m_2 < \ldots < m_{H+1}$$

such that $m_{H+1} - m_1 \ge H$ and, for every $j = 1, \ldots, n$,

$$H^{-1} > \left| \{m_{H+1}\alpha_j\} - \{m_1\alpha_j\} \right| = \left| (m_{H+1} - m_1)\alpha_j - \left(\left[m_{H+1}\alpha_j \right] - \left[m_1\alpha_j \right] \right) \right| .$$

We let $\widetilde{M} = m_{H+1} - m_1$ and, for every j, we let $p_j = \left[m_{H+1}\alpha_j \right] - \left[m_1\alpha_j \right]$. Then

$$\left| \widetilde{M}\alpha_j - p_j \right| < 1/H .$$

Since $H \le \widetilde{M} \le H^{n+1}$ we obtain (11.22). $\qquad\square$

The following result is due to Parnovski and Sobolev.

Theorem 11.9 (Parnovski and Sobolev) *Let $1 < d \equiv 1 \pmod 4$ and let $0 < r < 1/2$. For every $\varepsilon > 0$ there exists a diverging sequence $\{M_j\}_{j=1}^{+\infty}$ of positive integers satisfying*

$$D_{\mathrm{grid}}^2 \left(M_j^d \right) \le c_{\varepsilon, d, r} \, M_j^{d-1} \log^{-1/(d+\varepsilon)} \left(M_j \right) . \tag{11.23}$$

Proof For every positive integer j let

$$\widetilde{B}_j = \left\{ m \in \mathbb{Z}^d : 0 < |m| \le j^2 \right\} .$$

We have

$$\widetilde{B}_j \supset \left\{ m \in \mathbb{Z}^d : 0 < 2r|m| \le j^2 \right\} , \qquad \mathrm{card}\left(\widetilde{B}_j \right) \le 2^d j^{2d} .$$

Then Lemma 11.8 implies the existence of a diverging sequence M_j of positive integers satisfying

$$j \le M_j \le j^{2^d j^{2d}+1} , \qquad \left| \sin\left(2\pi r M_j |m| \right) \right| \le j^{-1} \tag{11.24}$$

for every $m \in \widetilde{B}_j$. The assumption $d \equiv 1 \pmod 4$ implies

$$\cos^2 \left(x - \frac{\nu\pi}{2} - \frac{\pi}{4} \right) = \sin^2(x)$$

when $\nu = d/2$ in (11.6). Let $N_j = M_j^d$ for every j. Then Parseval's identity, (11.18), (11.4) and (11.24) imply

$$D_{\mathrm{grid}}^2 \left(N_j \right) = N_j r^d \sum_{h \ne 0} |h|^{-d} J_{d/2}^2 \left(2\pi r M_j |h| \right) \tag{11.25}$$

$$= N_j r^d \sum_{0 < |h| \le j^2} |h|^{-d} J_{d/2}^2 \left(2\pi r M_j |h| \right) + N_j r^d \sum_{|h| > j^2} |h|^{-d} J_{d/2}^2 \left(2\pi r M_j |h| \right)$$

$$\leq M_j^{d-1} r^{d-1} \sum_{0<|h|\leq j^2} \pi^{-2} |h|^{-(d+1)} \sin^2\left(2\pi r M_j |h|\right)$$

$$+ M_j^{d-1} r^{d-1} \sum_{|h|>j^2} \pi^{-2} |h|^{-(d+1)} + O_d\left(M_j^{d-2} r^{d-2}\right)$$

$$\leq c_d \, M_j^{d-1} r^{d-1} j^{-2} \int_1^{j^2} s^{-2} \, ds + c \, M_j^{d-1} r^{d-1} \int_{j^2}^{+\infty} s^{-2} \, ds + O\left(M_j^{d-2} r^{d-2}\right)$$

$$\leq c_d' \, j^{-2} M_j^{d-1} r^{d-1} + O_d\left(M_j^{d-2} r^{d-2}\right) .$$

Observe that (11.24) implies, for every $\varepsilon > 0$,

$$\log\left(M_j\right) < \left(\left(2j^2\right)^d + 1\right)\log j < c_{\varepsilon,d} \left(2j^2\right)^{d+\varepsilon} .$$

Then the inequality

$$j^2 > c_{\varepsilon,d} \, \log^{1/(d+\varepsilon)}\left(M_j\right)$$

and (11.25) complete the proof. $\qquad\square$

Remark 11.10 The inequality (11.23) is almost sharp. Indeed, Parnovski and Sobolev [138, Theorem 3.1] have proved that if $d \equiv 1 \pmod 4$ then, for every positive real number δ, we have

$$D_{\mathrm{grid}}^2(N) \geq c_{d,\delta} \, M^{d-1-\delta} .$$

An inequality like (11.23) can be true also for sets different from balls. See [137, 175].

We can now prove the following result [51].

Theorem 11.11 *Let $d \equiv 1 \pmod 4$. Then for infinitely many values of M we have*

$$D_{\mathrm{grid}}^2\left(M^d\right) \leq \mathcal{D}_{\mathrm{jittered}}^2\left(M^d\right) .$$

Proof Let $N = M^d$. By Proposition 6.10 we have

$$\|\chi_{rB} * \lambda\|_{L^1(\mathbb{T}^d)} \leq \|\chi_{rB}\|_{L^1(\mathbb{T}^d)} \|\lambda\|_{L^1(\mathbb{T}^d)} = |rB| .$$

Then (11.13) and Corollary 6.6 yield

$$\mathcal{D}_{\mathrm{jittered}}^2(N) = N\left(|rB| - \|\chi_{rB} * \lambda\|_{L^2(\mathbb{T}^d)}^2\right)$$

$$= N\left(|rB| - \|\chi_{rB} * \lambda\|_{L^1(\mathbb{T}^d)}\right) + N\left(\|\chi_{rB} * \lambda\|_{L^1(\mathbb{T}^d)} - \|\chi_{rB} * \lambda\|_{L^2(\mathbb{T}^d)}^2\right)$$

$$\geq N \int_{\mathbb{T}^d}\left((\chi_{rB} * \lambda)(t) - (\chi_{rB} * \lambda)^2(t)\right) dt$$

$$= N \int_{\mathbb{T}^d} (\chi_{rB} * \lambda)(t)(1 - (\chi_{rB} * \lambda)(t)) \, dt \, .$$

Observe that

$$0 \le (\chi_{rB} * \lambda)(t)(1 - (\chi_{rB} * \lambda)(t)) \le 1 \qquad (11.26)$$

for every t. Also note that

$$(\chi_{rB} * \lambda)(t)(1 - (\chi_{rB} * \lambda)(t)) = 0$$

whenever.

$$\left(\left[-\frac{1}{2M}, \frac{1}{2M} \right]^d + t \right) \cap S(0, r) = \emptyset \, ,$$

where $S(0, r) = \{t : |t| = r\}$ is the sphere centred at 0, with radius r. We write $t = \rho\sigma$ in polar coordinates. For every $\sigma \in \Sigma_{d-1}$ (the unit sphere in $\mathbb{R}^d$) let $\mathcal{B}_{r,\sigma}$ be the half-space containing the origin and such that its boundary is tangent to the sphere $S(0, r)$ at the point $r\sigma$. By (11.16) and the symmetry of the cube about its centre we have

$$(\chi_{rB} * \lambda)(r\sigma) = N \left| rB \cap \left(\left[-\frac{1}{2M}, \frac{1}{2M} \right]^d + r\sigma \right) \right| = \frac{1}{2} + O_{d,r}\left(M^{-1} \right) \, .$$

By the monotonicity of the function $\rho \mapsto (\chi_{rB} * \lambda)(\rho\sigma)$ we have, for $r \le \rho \le r + 1/(4M)$,

$$0 \le (\chi_{rB} * \lambda)(\rho\sigma) \le \frac{1}{2} + O_{d,r}\left(M^{-1} \right) \, .$$

Then, for large M,

$$\mathcal{D}^2_{\text{jittered}}(N) \ge N \int_{r \le |t| \le r + 1/(4M)} (\chi_{rB} * \lambda)(t)(1 - (\chi_{rB} * \lambda)(t)) \, dt$$

$$\ge \frac{1}{3} N \int_{r \le |t| \le r + 1/(4M)} (\chi_{rB} * \lambda)(t) \, dt$$

$$\ge \frac{1}{3} N^2 \int_{r \le |t| \le r + 1/(4M)} \left| \mathcal{B}_{r,\sigma} \cap \left(\left[-\frac{1}{2M}, \frac{1}{2M} \right]^d + \rho\sigma \right) \right| \, dt + O_{d,r}\left(M^{d-2} \right)$$

$$\ge c_d N^2 \int_{r \le |t| \le r + 1/(4M)} M^{-d} \, dt + O_{d,r}\left(M^{d-2} \right) \ge c'_d M^{d-1} \, .$$

By appealing to (11.23) we complete the proof. $\qquad\qquad\square$

Exercises

1) Prove the existence of a positive constant c such that for every positive integer m and real $x > 0$, the Bessel function $J_m(x)$ satisfies

$$|J_m(x)| \le c\, x^{-1/3} \ .$$

2) Prove that for every non-negative integer k and every $x > 0$ we have

$$x^k \frac{d}{dx}\left(x^{-k} J_k(x)\right) = -J_{k+1}(x) \ .$$

3) Let $Q = [-1/2, 1/2]^d$. Prove that for every $\rho \ge 2$ we have

$$\int_{\Sigma_{d-1}} |\widehat{\chi}_Q(\rho\sigma)| \, d\sigma \ge c\, \frac{\log(\rho)}{\rho^d} \ .$$

4) Let rB be a d-dimensional ball of radius r, and let $\mathcal{D}_{\text{jittered}}$ be as in (11.10). Prove the existence of a positive constant $c_{d,r}$ such that

$$\frac{\mathcal{D}^2_{\text{jittered}}\left(M^d\right)}{M^{d-1}} \longrightarrow c_{d,r}$$

as $M \to +\infty$.

References

[1] T. van Aardenne-Ehrenfest, *Proof of the impossibility of a just distribution of an infinite sequence of points over an interval*, Proc. Kon. Ned. Akad. v. Wetensch **48** (1945), 266–271.

[2] T. van Aardenne-Ehrenfest, *On the impossibility of a just distribution*, Proc. Kon. Ned. Akad. v. Wetensch **52** (1949), 734–739.

[3] W.W. Adams, L.J. Goldstein, Introduction to number theory, Prentice-Hall, 1976.

[4] J. Agnew, Explorations in number theory, Contemporary Undergraduate Mathematics Series, Brooks/Cole, 1972.

[5] W.R. Alford, A. Granville, C. Pomerance, *There are infinitely many Carmichael numbers*, Ann. of Math. **139** (1994), 703–722.

[6] G.E. Andrews, Number theory, Dover Publications, 1994.

[7] G.E. Andrews, S.B. Ekhad, D. Zeilberger, *A Short Proof of Jacobi's formula for the number of representations of an integer as a sum of four squares*, Amer. Math. Monthly **100** (1993), 274–276.

[8] T.M. Apostol, Introduction to analytic number theory, Undergraduate Texts in Mathematics, Springer, 1998.

[9] A. Baker, A concise introduction to the theory of numbers, Cambridge University Press, 1984.

[10] P.T. Bateman, H.G. Diamond, Analytic number theory. An introductory course, World Scientific Publishing, 2004.

[11] D. Bayer, P. Diaconis, *Trailing the dovetail shuffle to its lair*, Ann. Appl. Probab. **2** (1992), 294–313.

[12] J. Beck, *Balanced two-colourings of finite sets in the square I*, Combinatorica **1** (1981), 50–64.

[13] J. Beck, *Irregularities of distribution I*, Acta Math. **159** (1987), 1–49.

[14] J. Beck, W.W.L. Chen, Irregularities of distribution, Cambridge Tracts in Mathematics, 89, Cambridge University Press, 2008.

[15] J. Beck, W.W.L. Chen, *Note on irregularities of distribution II*, Proc. London Math. Soc. **61** (1990), 251–272.

[16] D.R. Bellhouse, *Area estimation by point counting techniques*, Biometrics **37** (1981), 303–312.

[17] F. Benford, *The law of anomalous numbers*, Proc. Am. Philos. Soc. **78** (1938), 551–572.

[18] A. Berger, T.P. Hill, *Benford's law strikes back: no simple explanation in sight for mathematical gem*, Math. Intelligencer **33** (2011), 85–91.

[19] D. Bilyk, *Roth's orthogonal functions method in discrepancy theory and some new connections*, in 'A panorama of discrepancy theory' (W.W.L. Chen, A. Srivastav, G. Travaglini - Editors), Lecture Notes in Mathematics, Springer, to appear.

[20] S. Bochner, The role of mathematics in the rise of science, Princeton University Press, 1966.

[21] E. Borel, *Les probabilités dénombrables et leurs applications arithmétiques*, Rend. Circ. Mat. Palermo **27** (1909), 247–271.

[22] L. Brandolini, W.W.L. Chen, L. Colzani, G. Gigante, G. Travaglini, *Discrepancy and numerical integration in Sobolev spaces on metric measures spaces*, preprint.

[23] L. Brandolini, W.W.L. Chen, G. Gigante, G. Travaglini, *Discrepancy for randomized Riemann sums*, Proc. Amer. Math. Soc. **137** (2009), 3177–3185.

[24] L. Brandolini, C. Choirat, L. Colzani, G. Gigante, R. Seri, G. Travaglini, *Quadrature rules and distribution of points on manifolds*, Ann. Sc. Norm. Super. Pisa Cl. Sci., to appear

[25] L. Brandolini, L. Colzani, G. Gigante, G. Travaglini, *On the Koksma–Hlawka inequality*, J. Complexity **29** (2013), 158–172.

[26] L. Brandolini, L. Colzani, G. Gigante, G. Travaglini, *A Koksma–Hlawka inequality for simplices*, in 'Trends in harmonic analysis' (M. Picardello - Editor), Springer INdAM Series, Springer, 2013, 33–46.

[27] L. Brandolini, L. Colzani, A. Iosevich, A. Podkorytov, G. Travaglini, *Geometry of the Gauss map and lattice points in convex domains*, Mathematika **48** (2001), 107–117.

[28] L. Brandolini, L. Colzani, G. Travaglini, *Average decay of Fourier transforms and integer points in polyhedra*, Ark. Mat. **35** (1997), 253–275.

[29] L. Brandolini, G. Gigante, S. Thangavelu, G. Travaglini, *Convolution operators defined by singular measures on the motion group*, Indiana Univ. Math. J. **59** (2010), 1935–1945.

[30] L. Brandolini, G. Gigante, G. Travaglini, *Irregularities of distribution and average decay of Fourier transforms*, in 'A panorama of discrepancy theory' (W.W.L. Chen, A. Srivastav, G. Travaglini - Editors), Lecture Notes in Mathematics, Springer, to appear

[31] L. Brandolini, A. Greenleaf, G. Travaglini, $L^p - L^{p'}$ *estimates for overdetermined Radon transforms*, Trans. Amer. Math. Soc. **359** (2007), 2559–2575.

[32] L. Brandolini, S. Hofmann, A. Iosevich, *Sharp rate of average decay of the Fourier transform of a bounded set*, Geom. Funct. Anal. **13** (2003), 671–680.

[33] L. Brandolini, A. Iosevich, G. Travaglini, *Spherical means and the restriction phenomenon*, J. Fourier Anal. Appl. **7** (2001), 359–372.

[34] L. Brandolini, A. Iosevich, G. Travaglini, *Planar convex bodies, Fourier transform, lattice points, and irregularities of distribution*, Trans. Amer. Math. Soc. **355** (2003), 3513–3535.

[35] L. Brandolini, M. Rigoli, G. Travaglini, *Average decay of Fourier transforms and geometry of convex sets*, Rev. Mat. Iberoamer. **14** (1998), 519–560.

[36] L. Brandolini, G. Travaglini, *Pointwise convergence of Fejér type means*, Tohoku Math. J. **49** (1997), 323–336.

[37] L. Brandolini, G. Travaglini, *La legge di Benford*, Emmeciquadro **45** (2012).

[38] J. Bruna, A. Nagel, S. Wainger, *Convex hypersurfaces and Fourier transforms*, Ann. of Math. **127** (1988), 333–365.

[39] F. Cantelli, *Sulla probabilità come limite della frequenza*, Atti Accad. Naz. Lincei **26** (1917), 39–45.

[40] J.W.S. Cassels, *On the sums of powers of complex numbers*, Acta Math. Hungar. **7** (1957), 283–289.

[41] D.G. Champernowne, *The construction of decimal normal in the scale of ten*, J. London Math. Soc. **8** (1933), 254–260.

[42] K. Chandrasekharan, Introduction to analytic number theory, Die Grundlehren der mathematischen Wissenschaften, Band 148, Springer, 1968.

[43] B. Chazelle, The discrepancy method. Randomness and complexity, Cambridge University Press, 2000.

[44] W.W.L. Chen, *On irregularities of distribution III*, J. Austr. Math. Soc. **60** (1996), 228–244.

[45] W.W.L. Chen, Lectures on irregularities of point distribution, unpublished, 2000.

[46] W.W.L. Chen, Elementary number theory, unpublished, 2003.

[47] W.W.L. Chen, *Fourier techniques in the theory of irregularities of point distribution*, in 'Fourier analysis and convexity' (L. Brandolini, L. Colzani, A. Iosevich, G. Travaglini - Editors), Birkhauser, 2004, 59–82.

[48] W.W.L. Chen, M. Skriganov, *Upper bounds in irregularities of point distribution*, in 'A panorama of discrepancy theory' (W.W.L. Chen, A. Srivastav, G. Travaglini - Editors), Lecture Notes in Mathematics, Springer, to appear.

[49] W.W.L. Chen, A. Srivastav, G. Travaglini - Editors, A panorama of discrepancy theory, Lecture Notes in Mathematics, Springer, to appear.

[50] W.W.L. Chen, G. Travaglini, *Discrepancy with respect to convex polygons*, J. Complexity **23** (2007), 662–672.

[51] W.W.L. Chen, G. Travaglini, *Deterministic and probabilistic discrepancies*, Ark. Mat. **47** (2009), 273–293.

[52] W.W.L. Chen, G. Travaglini, *Some of Roth's ideas in discrepancy theory*, in 'Analytic number theory: essays in honour of Klaus Roth' (W.W.L. Chen, W.T. Gowers, H. Halberstam, W.M. Schmidt, R.C. Vaughan - Editors), Cambridge University Press, 2009, 150–163.

[53] P.R. Chernoff, *Pointwise convergence of Fourier series*, Amer. Math. Monthly **87** (1980), 399–400.

[54] K.L. Chung, A course in probability theory, Academic Press, 2001.

[55] M. Cipolla, *Sui numeri composti P che verificano la congruenza di Fermat $a^{P-1} \equiv 1 \pmod P$*, Ann. Mat. Pura Appl. **9** (1904), 139–160.

[56] J.A. Clarkson, *On the series of prime reciprocals*, Proc. Amer. Math. Soc. **17** (1966), 541.

[57] L. Colzani, G. Gigante, *Summation formulas and integer points under shifted generalized hyperbolae*, preprint.

232 References

[58] L. Colzani, G. Gigante, G. Travaglini, *Trigonometric approximation and a general form of the Erdős–Turán inequality*, Trans. Amer. Math. Soc. **363** (2011), 1101–1123.

[59] L. Colzani, G. Gigante, G. Travaglini, unpublished, 2012.

[60] L. Colzani, I. Rocco, G. Travaglini, *Quadratic estimates for the number of integer points in convex bodies*, Rend. Circ. Mat. Palermo **54** (2005), 241–252.

[61] J.H. Conway, R.K. Guy, The book of numbers, Copernicus, 1996.

[62] A.H. Copeland, P. Erdős, *Note on normal numbers*, Bull. Amer. Math. Soc. **52** (1946), 857–860.

[63] W.A. Coppel, Number theory. An introduction to mathematics, Springer, 2009.

[64] J.G. van der Corput, *Zalhentheorische abschätzungen*, Math. Ann. **84** (1921), 53–79.

[65] J.G. van der Corput, *Zalhentheorische abschätzungen mit anwendung auf gitterpunktprobleme*, Math. Z. **17** (1923), 250–259.

[66] J.G. van der Corput, *Verteilungsfunktionen I–VIII*, Proc. Akad. Amsterdam **38** (1935), 813–821, 1058–1066; **39** (1936), 10–19, 19–26, 149–153, 339–344, 489–494, 579–590.

[67] H. Davenport, *Notes on irregularities of distribution*, Mathematika **3** (1956), 131–135.

[68] J. De Koninck, F. Luca, Analytic number theory. Exploring the anatomy of integers. Graduate Studies in Mathematics, 134, American Mathematical Society, 2012.

[69] P. Diaconis, *The distribution of leading digits and uniform distribution* mod 1, Ann. Prob. **5** (1977), 72–81.

[70] P. Diaconis, D. Freedman, *On rounding percentages*, J. Amer. Statist. Assoc. **366** (1979), 359–364.

[71] J. Dick, F. Pillichshammer, *Discrepancy theory and quasi-Monte Carlo integration*, in 'A panorama of discrepancy theory' (W.W.L. Chen, A. Srivastav, G. Travaglini - Editors), Lecture Notes in Mathematics, Springer, to appear.

[72] L.E. Dickson, History of the theory of numbers, Vol. I, II, Chelsea Publishing Co., 1966.

[73] M. Drmota, R.F. Tichy, Sequences, discrepancies and applications. Lecture Notes in Mathematics, 1651, Springer, 1997.

[74] P. Erdős, *On almost primes*, Amer. Math. Monthly **57** (1950), 404–407.

[75] P. Erdős, W.H.J. Fuchs, *On a problem of additive number theory*, J. London Math. Soc. **31** (1956), 67–73.

[76] P. Erdős, J. Suranyi, Topics in the theory of numbers, Undergraduate Texts in Mathematics, Springer, 2003.

[77] P. Erdős, P. Turán, *On a problem in the theory of uniform distribution I, II*, Indag. Math. **10** (1948), 370–378, 406–413.

[78] G. Everest, T. Ward, An introduction to number theory, Graduate Texts in Mathematics, 232, Springer, 2005.

[79] D.E. Flath, Introduction to number theory, John Wiley & Sons, 1989.

[80] B. Flehinger, *On the probability that a random integer has initial digit A*, Amer. Math. Monthly **73** (1966), 1056–1061.

[81] G.B. Folland, Fourier analysis and its applications, Wadsworth & Brooks/Cole, 1992.

[82] G.B. Folland, Real analysis. Modern techniques and their applications, John Wiley & Sons, 1999.

[83] L.J. Goldstein, *A history of the prime number theorem*, Amer. Math. Monthly **80** (1973), 599–615.

[84] S.W. Graham, G. Kolesnik, Van der Corput's method of exponential sums, London Mathematical Society Lecture Note Series, 126, Cambridge University Press, 1991.

[85] A. Granville, *The Fundamental theorem of arithmetic*, preprint.

[86] A. Granville, Z. Rudnick - Editors, Equidistribution in number theory, an introduction, Springer, 2007

[87] T.H. Gronwall, *Some asymptotic expressions in the theory of numbers*, Trans. Amer. Math. Soc. **14** (1913), 113–122.

[88] G.H. Hardy, *On the expression of a number as the sum of two squares*, Quart. J. Math. **46** (1915), 263–283.

[89] G.H. Hardy, *On Dirichlet's divisor problem*, Proc. London Math. Soc. **15** (1916), 1–25.

[90] G.H. Hardy, E.M. Wright, An introduction to the theory of numbers, Oxford University Press, 1938.

[91] G. Harman, Metric number theory, London Mathematical Society Monographs, New Series, 18, Oxford University Press, 1998.

[92] G. Harman, *Variations on the Koksma–Hlawka inequality*, Unif. Distr. Theory **5** (2010), 65–78.

[93] H. Hasse, Number theory, Springer, 1980.

[94] F.J. Hickernell, *Koksma–Hlawka inequality*, in 'Encyclopedia of statistical sciences' (S. Kotz, C.B. Read, D.L. Banks - Editors), Wiley-Interscience, 2006.

[95] E. Hlawka, The theory of uniform distribution, AB Academic Publishers, 1984.

[96] E. Hlawka, J. Schoißengeier, R. Taschner, Geometric and analytic number theory, Universitext, Springer, 1991.

[97] P. Hoffman, The man who loved only numbers: The story of Paul Erdős and the search for mathematical truth, Hyperion Books, 1998.

[98] L.K. Hua, Introduction to number theory, Springer, 1982.

[99] M.N. Huxley, *The mean lattice point discrepancy*, Proc. Edinburgh Math. Soc. **38** (1995), 523–531.

[100] M.N. Huxley, Area, lattice points and exponential sums, London Mathematical Society Monographs, New Series, 13, Oxford Science Publications, 1996.

[101] K. Ireland, M. Rosen, A classical introduction to modern number theory, Graduate Texts in Mathematics, 84, Springer, 1990.

[102] A. Ivić, The Riemann zeta-function. Theory and applications, Dover Publications, 2003.

[103] G.A. Jones, J.M. Jones, Elementary number theory, Springer Undergraduate Mathematics Series, Springer, 1998.

[104] C. Joy, P.P. Boyle, K.S. Tan, *Quasi-Monte Carlo methods in finance*, Management Science **42** (1996), 926–938.

[105] Y. Katznelson, An introduction to harmonic analysis, Cambridge Mathematical Library, Cambridge University Press, 2004.

[106] D.G. Kendall, *On the number of lattice points in a random oval*, Quart. J. Math. Oxford Series **19** (1948), 1–26.

[107] N. Koblitz, A course in number theory and cryptography, Graduate Texts in Mathematics, 114, Springer, 1994.

[108] H. Koch, Number theory, Graduate Studies in Mathematics, American Mathematical Society, 2000.

[109] J.F. Koksma, *Een algemeene stellinguit de theorie der gelijkmatige verdeeling modulo* 1, Mathematica B (Zupten) **11** (1942-43), 7–11.

[110] T. Kollig, A. Keller, *Efficient multidimensional sampling*, Computer Graphics Forum **21** (2002), 557–563.

[111] M.N. Kolountzakis, T. Wolff, *On the Steinhaus tiling problem*, Mathematika **46** (1999), 253–280.

[112] S.V. Konyagin, M.M. Skriganov, A.V. Sobolev, *On a lattice point problem arising in the spectral analysis of periodic operators*, Mathematika **50** (2003), 87–98.

[113] E. Kratzel, Lattice points, Mathematics and its Applications, Kluwer Academic Publisher, 1988.

[114] L. Kuipers, H. Niederreiter, Uniform distribution of sequences, Dover Publications, 2006.

[115] E. Landau, *Über die gitterpunkte in einen kreise (Erste, zweite Mitteilung)*, Nachr. K. Gesellschaft Wiss. Göttingen, Math.-Phys. Klasse (1915), 148–160, 161–171.

[116] E. Landau, ber Dirichlets teilerproblem, Sitzungsber, Math.–Phys. Klasse Knigl. Bayer. Akad. Wiss. (1915), 317–328.

[117] N.N. Lebedev, Special functions and their applications, Dover Publications, 1972.

[118] F. Lemmermeyer, Reciprocity laws (from Euler to Eisenstein), Springer Monographs in Mathematics, Springer, 2000.

[119] W.J. LeVeque, Fundamentals of number theory, Addison-Wesley, 1977.

[120] W.J. LeVeque, Elementary theory of numbers, Dover Publications, 1990.

[121] J. Matousek, Geometric discrepancy. An illustrated guide, Algorithms and Combinatorics, 18, Springer, 2010.

[122] R. Matthews, *The power of one*, New Scientist, 10 July 1999.

[123] H. Montgomery, Ten lectures on the interface between analytic number theory and harmonic analysis, CBMS Regional Conference Series in Mathematics, 84, American Mathematical Society, 1994.

[124] W.J. Morokoff, R.E. Caflisch, *A quasi-Monte Carlo approach to particle simulation of the heat equation*, SIAM J. Numer. Anal. **30** (1993), 1558–1573.

[125] R.M. Murty, N. Thain, *Prime numbers in certain arithmetic progressions*, Funct. Approx. Comment. Math. **35** (2006), 249–259.

[126] R.M. Murty, N. Thain, *Pick's theorem via Minkowski's theorem*, Amer. Math. Monthly **114** (2007), 732–736.

[127] W. Narkiewicz, Number theory, World Scientific, 1983.

[128] M.B. Nathanson, Elementary methods in number theory, Graduate Texts in Mathematics, 195, Springer, 2000.

[129] S. Newcomb, *Note on the frequency of use of the different digits in natural numbers*, Amer. J. Math. **4** (1881), 39–40.

[130] D.J. Newman, *A simplified proof of the Erdős–Fuchs theorem*, Proc. Amer. Math. Soc. **75** (1979), 209–210.

[131] H. Niederreiter, Random number generation and quasi-Monte Carlo methods, CBMS-NSF Regional Conference Series in Applied Mathematics, 63, SIAM, 1992.

[132] M. Nigrini, Benford's law. Applications for forensic accounting, auditing, and fraud detection, John Wiley & Sons, 2012.

[133] M. Nigrini, L. Mittermaier, *The use of Benford's law as an aid in analytical procedures*, Auditing - A Journal of Practice & Theory **16** (1997), 52–67.

[134] I. Niven, Irrational numbers, Carus Mathematical Monographs, 11, MAA, 2005.

[135] I. Niven, H. Zuckerman, An introduction to the theory of numbers, John Wiley & Sons, 1980.

[136] O. Ore, Number theory and its history, Dover Publications, 1988.

[137] L. Parnovski, N. Sidorova, *Critical dimensions for counting lattice points in Euclidean annuli*. Math. Model. Nat. Phenom. **5** (2010), 293–316.

[138] L. Parnovski, A. Sobolev, *On the Bethe–Sommerfeld conjecture for the polyharmonic operator*, Duke Math. J. **107** (2001), 209–238.

[139] R. Pinkham, *On the distribution of first significant digits*, Ann. Math. Stat. **32** (1961), 1223–1230.

[140] M.A. Pinsky, Introduction to Fourier analysis and wavelets, Graduate Studies in Mathematics, 102, American Mathematical Society, 2002.

[141] M. Plancherel, *Contribution a l'etude de la representation d'une fonction arbitraire par les integrales définies*, Rend. del Circ. Mat. Palermo **30** (1910), 298–335.

[142] A.N. Podkorytov, *The asymptotic of a Fourier transform on a convex curve,* Vestn. Leningr. Univ. Mat. **24** (1991), 57–65.

[143] S. Ramanujan, *A proof of Bertrand's postulate*, J. Indian Math. Soc. **11** (1919), 181–182.

[144] B. Randol, *On the Fourier transform of the indicator function of a planar set,* Trans. Amer. Math. Soc. **139**, 271–278.

[145] D. Redmond, Number theory. An introduction, Monographs and Textbooks in Pure and Applied Mathematics, 201, Marcel Dekker, 1996.

[146] E. Regazzini, *Probability and statistics in Italy during the First World War I: Cantelli and the laws of large numbers*, J. Électron. Hist. Probab. Stat. **1** (2005) 1–12.

[147] F. Ricci, G. Travaglini, *Convex curves, Radon transforms and convolution operators defined by singular measures*, Proc. Amer. Math. Soc. **129** (2001), 1739–1744.

[148] S. Robinson, *Still guarding secrets after years of attacks, RSA earns accolades for its founders*, SIAM News **36** 5 (2003).

[149] K.F. Roth, *On irregularities of distribution,* Mathematika **1** (1954), 73–79.

[150] W. Rudin, Principles of mathematical analysis, International Series in Pure and Applied Mathematics, McGraw-Hill, 1976.

[151] J.D. Sally, P.J. Sally, Roots to research. A vertical development of mathematical problems, American Mathematical Society, 2007.

[152] W.M. Schmidt, *Irregularities of distribution IV*, Invent. Math. **7** (1968), 55–82.

[153] W.M. Schmidt, Lectures on irregularities of distribution, Tata Institute of Fundamental Research Lectures on Mathematics and Physics, 56, Tata Institute of Fundamental Research, Bombay, 1977.

[154] E. Scholz (Editor), Hermann Weyl's Raum–Zeit–Materie and a general introduction to his scientific work, DMV Seminar 30, Birkhauser, 2001.

[155] W. Sierpinski, *O Pewnem zagadnieniu z rachunku funckcyj asymptotycznych,* Prace mat.-fiz. **17** (1906), 77–118.

[156] R.A. Silverman, Complex analysis with applications, Dover Publications, 1984.

[157] S. Singh, The code book: The science of secrecy from ancient Egypt to quantum cryptography, Doubleday Books, 1999.

[158] I.H. Sloan, H. Wozniakowski, *When are quasi-Monte Carlo algorithms efficient for high dimensional integrals?* J. Complexity **14** (1998), 1–33.

[159] P. Soardi, Serie di Fourier in più variabili, Unione Matematica Italiana – Pitagora, Quaderni dell'Unione Matematica Italiana 26, 1984.

[160] C.D. Sogge, Fourier integrals in classical analysis, Cambridge Tracts in Mathematics, 105, Cambridge University Press, 1993.

[161] E. Stein, R. Shakarchi, Fourier analysis, An introduction, Princeton Lectures in Analysis, I, Princeton University Press, 2003.

[162] E. Stein, R. Shakarchi, Real analysis, measure theory, integration, and Hilbert spaces, Princeton Lectures in Analysis, III, Princeton University Press, 2005.

[163] E. Stein, R. Shakarchi, Functional analysis, Princeton Lectures in Analysis IV, Princeton University Press, 2011.

[164] E. Stein, G. Weiss, Introduction to Fourier analysis in Euclidean spaces, Princeton Mathematical Series, 32, Princeton University Press, 1971.

[165] I.N. Stewart, D.O. Tall, Algebraic number theory, Chapman and Hall Mathematics Series, Chapman and Hall, 1979.

[166] J. Stillwell, Elements of algebra, geometry, numbers, equations, Undergraduate Texts in Mathematics, Springer, 1994.

[167] J. Stillwell, Elements of number theory, Undergraduate Texts in Mathematics, Springer, 2003.

[168] D.R. Stinson, Cryptography, theory and practice, CRC Press Series on Discrete Mathematics and its Applications, Chapman & Hall/CRC, 2002.

[169] M. Tarnopolska-Weiss, *On the number of lattice points in planar domains,* Proc. Amer. Math. Soc. **69** (1978), 308–311.

[170] G. Travaglini, *Fejer kernels for Fourier series on* $\mathbb{T}^n$ *and on compact Lie groups,* Math. Z. **216** (1994), 265–281.

[171] G. Travaglini, *Crittografia,* Emmeciquadro **21** (2004), 21–28.

[172] G. Travaglini, *Average decay of the Fourier transform,* in 'Fourier analysis and convexity' (L. Brandolini, L. Colzani, A. Iosevich, G. Travaglini - Editors), Birkhauser, 2004, 245–268.

[173] G. Travaglini, Appunti su teoria dei numeri, analisi di Fourier e distribuzione di punti, Unione Matematica Italiana – Pitagora, Quaderni dell'Unione Matematica Italiana 52, 2010.

[174] K. Tsang, *Recent progress on the Dirichlet divisor problem and the mean square of the Riemann zeta-function,* Sci. China Math. **53** (2010), 2561–2572.

[175] M. Tupputi, in preparation.

[176] J.D. Vaaler, *Some extremal problems in Fourier analysis,* Bull. Amer. Math. Soc. **12** (1985), 183–216.

[177] G. Voronoï, *Sur un problème du calcul des fonctions asymptotiques,* J. Reine Angew. Math. **126** (1903), 241–282.

[178] G.N. Watson, A treatise on the theory of Bessel functions, Cambridge Mathematical Library, Cambridge University Press, 1922.

[179] D.D. Wall, Normal numbers, Thesis, University of California, 1949.

[180] H. Weyl, *Uber ein problem aus dem gebiete der diophantischen approximationen*, Nacr. Ges. Wiss. Gottingen (1914), 234–244.

[181] H. Weyl, *Uber die gleichverteilung von zhalen mod. eins*, Math. Ann. **77** (1916), 313–352.

[182] R.L. Wheeden, A. Zygmund, Measure and integral. An introduction to real analysis, Pure and Applied Mathematics, 43, Marcel Dekker, 1977.

[183] Y. Zhang, *Bounded gaps between primes*, Ann. Math. **179** (2014), 1121–1174.

[184] G. Ziegler, *The great prime number record races*, Notices Amer. Math. Soc. **51** (2004), 414–416.

[185] A. Zygmund, Trigonometric series, I, II, Cambridge Mathematical Library, Cambridge University Press, 1993.

Index

Printed in the United States
By Bookmasters